SpringerBriefs in Cybersecurity

Editor-in-Chief

Sandro Gaycken, Digital Society Institute
European School of Management and Technology (ESMT)
Stuttgart, Baden-Württemberg, Germany

Series Editors

Sylvia Kierkegaard, International Association of IT Lawyers
Highfield, Southampton, UK

John Mallery, Computer Science and Artificial Intelligence,
Massachusetts Institute of Technology
Cambridge, MA, USA

Steven J. Murdoch, University College London
London, UK

Kenneth Geers, Taras Shevchenko University
Kyiv, Kievs'ka, Ukraine

Michael Kasper, Department of Cyber-Physical Systems Security
Fraunhofer Institute SIT
Darmstadt, Hessen, Germany

Cybersecurity is a difficult and complex field. The technical, political and legal questions surrounding it are complicated, often stretching a spectrum of diverse technologies, varying legal bodies, different political ideas and responsibilities. Cybersecurity is intrinsically interdisciplinary, and most activities in one field immediately affect the others. Technologies and techniques, strategies and tactics, motives and ideologies, rules and laws, institutions and industries, power and money – all of these topics have a role to play in cybersecurity, and all of these are tightly interwoven.

The *SpringerBriefs in Cybersecurity* series is comprised of two types of briefs: topic- and country-specific briefs. Topic-specific briefs strive to provide a comprehensive coverage of the whole range of topics surrounding cybersecurity, combining whenever possible legal, ethical, social, political and technical issues. Authors with diverse backgrounds explain their motivation, their mindset, and their approach to the topic, to illuminate its theoretical foundations, the practical nuts and bolts and its past, present and future. Country-specific briefs cover national perceptions and strategies, with officials and national authorities explaining the background, the leading thoughts and interests behind the official statements, to foster a more informed international dialogue.

Luigi Martino

Cybersecurity in Italy

Governance, Policies and Ecosystem

 Springer

Luigi Martino
University of Bologna
Bologna, Italy

Khalifa University of Science and Technology
Abu Dhabi, UAE

ISSN 2193-973X   ISSN 2193-9748   (electronic)
SpringerBriefs in Cybersecurity
ISBN 978-3-031-64395-8   ISBN 978-3-031-64396-5   (eBook)
https://doi.org/10.1007/978-3-031-64396-5

This Springer imprint is published by the registered company Springer Nature Switzerland AG
The registered company address is: Gewerbestrasse 11, 6330 Cham, Switzerland

If disposing of this product, please recycle the paper.

"To my beloved wife Rugiada"

"We do raise to high degrees of knowledge whomever We will—but above everyone who is endowed with knowledge there is One who knows all"

Foreword

I first met Luigi Martino in 2018 when he came, as a visiting PhD student, to spend time with our team at University College London. Luigi was supervised by Giampiero Giacomello, an academic I greatly admired who had examined my own PhD at the Australian National University. Having Luigi spend time in our team was a concrete way to continue to build links within that intellectual community of scholars working at the intersection of emerging digital technologies and international relations. Luigi would later go on to head the Center for Cyber Security and International Relations Studies at Firenze which many of us have now visited and spent time at.

Luigi's participation in the diplomacy of global cybersecurity has provided him with a special perspective that views the national interest and multilateral practice as interdependent. We cannot understand discourse on the state of global affairs and cybersecurity without deep engagement with state level approaches to the legal, regulatory, governance and policy issues. Ultimately, cybersecurity has such profound and interconnected economic, social, political, and industrial implications that understanding domestic factors and drivers is essential to understanding the international ecosystem in which those implications play out.

Fundamentally, Luigi's work highlights that cybersecurity as it is not only a problem to be addressed but also an enabler to be maximized. There are opportunity costs of not implementing cybersecurity that ripple through the ambitions and expectations of states and threaten to hold back extraordinary progress that might otherwise be open to us.

Cybersecurity is a classic "wicked policy problem" in which initiatives that solve one problem can simultaneously create others. As Luigi's book documents, cybersecurity policy and governance are not simply a matter of "evidence-based policy making"—this is a holistic and all-encompassing challenge that involves a multitude of actors, systems, interests, and values.

This book is an important counterweight to the dominant narrative of the past three decades that "governments stifle innovation." As we now recognize that market drivers alone have not, and will not, deliver cybersecurity in the manner, and on the scale, needed by society, works like this make a real contribution. The conversation must now turn to *how* governments can best ensure that cybersecurity outcomes are aligned to the broader national interest. This book carefully and systematically details how Italy has approached the design and implementation of policy initiatives, regulations, incentives, and guidance to balance out the complex array of factors that shape, and are influenced by, cybersecurity.

The issue areas that Luigi selects for his study are an important indicator of his perspective and of how one might engage with the book. Certainly, there is a clear logic that runs through from the Introduction and his thoughts on the intersection of politics and cyberspace, right through to the Conclusion where he offers his views on the future of Italian cybersecurity governance. I read this book in this linear fashion and found that the narrative builds with each chapter to provide a really comprehensive understanding of the Italian cyberspace ecosystem. However, many of the chapters stand alone and a reader could certainly engage with the book in a non-linear, issues-driven approach. I found the historical elements of these chapters particularly compelling, as they provided a wider, socio-political context for the Italian view. But one could also focus on the regulatory chapter or the institutional chapter in a comparative case study.

The analysis of the Italian history and contemporary approach to cybersecurity governance that Luigi provides here will be an excellent resource for students, certainly. But also, for those working on similar problems in other government, private sector, and third sector settings who can learn from, compare with, and evaluate the Italian response to this epoch defining public policy challenge. Perhaps even more significantly, Luigi's work provides a clear, analytic framework for research, understanding, and comparing other state responses. This book is certain to be influential and impactful and I congratulate Luigi on his achievement and thank him for his contribution to the literature.

London, UK
2024   Madeline Carr

Preface

This book offers a focused perspective on the governance, policies, legal frameworks, ecosystem, and national architecture implemented by Italy in the context of cybersecurity. Given Italy's significant geopolitical position, and its status as a mature country in the digital society, the comprehensive exploration included in this book attempts to shed light on the nuanced ways in which the nation addresses digital threats, adapts to technological advancements, and applies laws that protect and enhance the well-being of its citizens and organizations. For professionals working in Italy or with Italy, this detailed analysis is immediately applicable. Understanding Italy's governance structures, therefore, helps cybersecurity professionals navigate a complex institutional (and bureaucratic) framework consisting of stakeholders and decision-makers, thereby improving their knowledge of the roles that require interactions with government entities or compliance with national policies and rules. This book also serves as an essential reference for those involved in policymaking or managing cybersecurity from both a public and private perspective.

However, this book is dedicated specifically to students who are interested in, or want to become involved in, the political aspect of cybersecurity. In this book, they will hopefully find valuable information on "who" does "what" in Italy in the context of cybersecurity. By studying the Italian approach, students can appreciate the practical implications of theoretical knowledge, such as the implementation of laws in the field, how policies are shaped in response to emerging threats, and how governance structures influence national cybersecurity strategies. Designed to promote a critical thinking approach, and based on empirical validation of data with 35 semi-structured interviews, this book encourages students to analyze and question "how" and "why" certain decisions are made in the context of national cybersecurity. Addressing a significant gap in the existing literature, this book provides an updated and in-depth analysis of Italy's response to cybersecurity issues, setting a precedent for similar studies in other national contexts. As cybersecurity assumes a greater role in national security, the insights gleaned from Italy's experiences hold relevance not only for those within Italy studying or working in this field, but also for the global community endeavoring to fortify its cybersecurity measures. Moreover, the book offers valuable perspectives for individuals seeking to comprehend the strategic stance of a specific country in this field. By presenting a holistic view of the Italian cybersecurity framework, this book also aims to contribute to the dissemination of a culture of cybersecurity awareness in Italy. It highlights the significance of comprehending the Italian cybersecurity model, while also recognizing that Italy, within the cybersecurity landscape, is a component of a broader framework encompassing its role as a Member State of the European Union and its position as a significant actor globally in different international (i.e. UN; G20 and G7 etc.) and regional (i.e. OSCE, etc.) fora.

This book represents the culmination of extensive research, more than 15 years of personal experiences, academic research and professional insights. However, it is important to note that the contents provided here are based on the author's interpretation and the content represents a snapshot of the subject at the time of writing. As the author of this book, I hereby assume full responsibility for the contents provided herein. While every effort has been made to ensure the accuracy, reliability, and completeness of the information presented, I recognize that I am solely responsible for any errors, inaccuracies, or omissions that may be found.

Bologna, Italy   Luigi Martino
June, 2024

Acknowledgments

I am profoundly grateful to several individuals whose invaluable contributions have shaped the essence of this book. I am deeply indebted to my mentor, Umberto Gori, a pioneering scholar who has masterfully applied classical theories of Security Studies and International Relations to analyze the phenomena of cyberspace. His futuristic wisdom, profound insights, and rigorous approach have significantly influenced the international debate on this topic. His active contributions have both anticipated and shaped our understanding and the methodologies we use to conduct research in this field. His mentorship has not only sculpted my academic journey, but also equipped me to navigate the complex dynamics of cyberspace with a policy- and security-oriented approach. Special thanks to Giampiero Giacomello for his active contributions to the innovation of our discipline with his cutting-edge research and for his dissemination at international level. Thanks also to Francesco N. Moro, whose persistent encouragement was instrumental in the publication of this book, and to Giorgio Natalicchi for his contributions to my academic growth. Ernesto Damiani deserves special mention for his invaluable support and pragmatic advice, which have not only enhanced the content of this book, but also contributed directly to the maturity of both the Italian and international cybersecurity community. Thanks to Marco Mayer, one of the most visionary policymakers and researchers I have encountered. I must also extend my gratitude to several colleagues. Madeline Carr for accepting the responsibility of including her foreword in this book, for the support during my PhD and for her trust in me. Thanks to Chris Painter, Bushra AlBlooshi, Shujun Li, Charity Wright, John Mallery, Andrea Calderaro, Antonio Villafranca, Nazli Choucri, Joseph Nye, Alessandro Gili, Lior Tabansky, Lauren Zabierek, Chris Spirito and Fabio Rugge. I am grateful to Massimo Mercati for our genuine friendship, and to Marco Lisi for his active contributions to making Space a more cyber-secure environment. A special thanks to Raffaele Boccardo for his valuable approach based on a genuine mentorship; and to Luciano Bozzo and Adriano Soi for their guidance. Samuele Foni deserve a special mention, because he exemplifies how expertise and friendship can combine into a lifelong brotherhood. Rocco Mammoliti has been a major influence in exchanging visions on the need for protecting Italy's cybersecurity critical infrastructure. Michele Colajanni's unwavering belief in the significance of applying a social sciences perspective to the computer science field has been a constant source of motivation. I would like to thank all associates of the Center for Cyber Security and International Relations Studies, whose direct and indirect contributions have been crucial in shaping the contents of this book. A special mention goes to Valentina Luna Covella for her meticulous assistance in processing the figures and tables. Nadeen Gamal's unwavering support, editing prowess, and invaluable assistance have been indispensable throughout this journey. Additional thanks go to those involved in the Italian cybersecurity environment, whose daily efforts make the country a frontrunner in this sector. I am grateful to the 35 interviewees who, while with their choice to remain anonymous, they are committed to work behind the scenes and have contributed significantly to this book. Their approach is committed to contributing systematically to the success of Italy, echoing the words of one of my mentors: "If we want to receive applause, we have to work at the circus". Last but not least, special thanks to my family, starting with my beloved wife, for her patience, advice and to be my role model. Thanks also to my "extended" family. This book represents the response to those who advocate for "using knowledge for good".

Contents

List of Figures

List of Graphs

List of Tables

Chapter 1
Cybersecurity: Intersecting Technology and Policy

1.1 Introduction

The expansion of cyberspace has marked a clear distinction between the time preceding and following the onset of the digital era. As in every stage of human evolution and the revolutions associated with it, the onset of the information technology revolution has sparked a gradual blending between the technological realm and the crafting of suitable policies for the cyber phenomenon. From its inception, this phenomenon has displayed social implications that stretch beyond mere technical aspects. The intersection of the technological domain and policy has been driven by the need to address an inherent imbalance in the development of the information technology context, namely, the evidence that innovations in the ICT domain tend to prioritize efficiency over security. To probe into this evidence, we can cite Weinberg's law which states that "If builders built buildings the way programmers write programs, then the first woodpecker that came along would destroy civilization" (Chemuturi 2010, ix). This simple yet pragmatic observation is also closely related to an inherent paradox of digitization: a trade-off where increased computerization implies decreased security. Indeed, the implications arising from the expansion of technological innovations, especially in the information technology field, have shown the potential to negatively impact national security, the security of citizens, the realm of human rights, and, in other words, everything on which humans base their activities. Therefore, it was necessary to adopt a policy-oriented approach through the creation of a regulatory framework that, beyond the sector standards, aimed to govern the cyber phenomenon. In this way, starting from the late 1990s and the early years of the new millennium, there was a decision to overcome the myopic barrier initially built by IT professionals but, to some extent, also fueled by the reluctance of policymakers who adopted the axiom that cyber dynamics "because exclusively of an informational nature, hence technical, and thus nonpolitical, are the stuff of engineers or nerds" (Interview, Rome 2024).

L. Martino, *Cybersecurity in Italy*, SpringerBriefs in Cybersecurity,
https://doi.org/10.1007/978-3-031-64396-5_1

This barrier, entirely myopic and at times ideological, has been shattered by the awareness that in the cyberspace context, four relevant factors converge to build a cybersecurity based on the intersection between policy and technology: (1) the exponential growth of dependence on ICT systems for vital functions of modern societies, (2) the inherent presence of vulnerabilities in the development of ICT technologies and the consequent cause of malfunction, (3) the human factor understood as a point of vulnerability and susceptibility to criminal activities, and (4) the tendency of state and non-state actors to use ICT tools for political purposes with the risk of military escalation.

Hence, it follows the need to create a suitable security model to mitigate the risks and threats arising from cyber dynamics, which has both a technical and political components. The former is capable of addressing the "how" of securing ICT tools and the environment in which they operate from attacks or human errors. The latter can address the need for "how" to ensure that malicious dynamics produced in and through cyberspace can be limited, reduced, or eliminated, preserving social, national, and economic security and identifying at the governmental level "who" is in charge of doing so.

The concept of cybersecurity is therefore to be understood in its dual scope, both as a technical practice and as a sociopolitical necessity, to address how policy and technology can confront the threats of the cyber world. Seen from both a technical and a political perspective, cybersecurity includes measures and practices aimed at protecting computer systems, networks, and digital information from a wide range of threats, vulnerabilities, and attacks.

From a technical perspective, cybersecurity focuses on safeguarding the confidentiality, integrity, and availability of digital resources, information, and data. This technical perspective is accompanied by an analysis and intervention level of a political-administrative nature. From this standpoint, cybersecurity assumes broader implications as it concerns the protection of a nation's critical infrastructure, military resources, economic interests, and the general well-being of its citizens. In summary, cybersecurity should be understood as a dynamic approach oriented toward the management of technological processes and inclusive of policies aimed at preventing, countering, and mitigating the negative effects of events produced in and through cyberspace, including those with the aim of compromising national security. This process must be flexible enough to accommodate technological progress and evolving threats while considering the capabilities of malicious actors and the multifaceted nature of cyber challenges. It encompasses not only safeguarding digital assets but also protecting national interests, security, and the welfare of a nation and its citizens in an interconnected and digitally reliant world.

Implications of Political Dynamics in Cyberspace

Cyberspace has simultaneously become the content, container, and medium for social, economic, political, military, symbolic, and technological activities in contemporary times. From an environmental viewpoint, cyberspace possesses peculiar characteristics such as "placelessness," ubiquity, and anonymity (Gray 2013, 15), but these features do not exempt it from being included in the classic conceptual framework of politics. Similar to other environments where human activities unfold, cyberspace is also affected by conflicts, negotiations, and other political mechanisms (Choucri 2012; Valeriano and Maness 2014). While literature defines cyberspace as an artificial environment, in practical terms, its configuration is based on a hybrid interaction between geographic, physical, logical-virtual, social, and environmental components. This combination of physical and virtual elements implies that cyberspace is understood as a resource and an arena for political action in the twenty-first century. For instance, the physical layer of cyberspace (commonly understood as the combination of hardware and infrastructure such as terrestrial and submarine cables, satellites, antennas, routers, computers, servers, data centers, etc.) must be connected with the geographical dimension, i.e., the territory where the physical components are materially installed. Consequently, there is a political implication arising from the interaction between those governing the territory and expressing their jurisdiction and those who own or manage the technological infrastructure installed in that specific territory.

However, the virtual feature of cyberspace contributes to challenging certain axioms that were previously taken for granted and immutable in the context of politics before the digital era. Among these implications, it is worth mentioning the erosion of the role of the nation-state in relation to the monopoly of violence and information, the speed of action execution, the low barrier of access to technology, and the implications of anonymity.

Focusing on the first implication, the advent of cyberspace has marked a revision of the concept of information sovereignty and the monopoly of internal and external violence of nation-states. In particular, the concepts of the monopoly of violence and information, classically understood as fundamental elements of the political power of state actors and government bureaucracies, are now shared and partly held by non-state actors who possess greater capabilities in terms of technological, human, and economic resources compared to states. Since these non-state actors control a share of essential resources for the well-being of contemporary societies, they also have the ability to impose rules aimed at protecting their own interests, which in some cases conflict with those of states.

A practical example of this situation is the role of non-state actors that own social platforms, where communication and social interaction are expressed today. The rules of terms and conditions for accessing, sharing content, managing data, and interacting on these platforms are dictated not by governments but by providers, based on their interests. Although it may seem like an obvious and simplistic fact, the implications become evident when understanding the extent of "who" holds the power to decide "what" to share and "how" to access such information.

Another political element wielded by cyberspace is the overshadowing of the concept of "trust" in the context of international relations, due to the use of cyber capabilities to achieve political and military objectives. In cyberspace, the fundamental principles of peace and stability in the international system, such as the assurance that agreements are honored and transparency in identifying allies and adversaries (Bull 1977), are called into question. The certainty of identifying the perpetrator of a malicious act is subject to plausible deniability (Eriksson & Giacomello 2006; Dunn-Cavelty 2008; Carr 2017; Eriksson & Giacomello 2023).

Anonymity tends to generate three conditions that push toward a vicious circle: (a) an increase in the perception of danger with a consequent security dilemma, (b) a rush toward the militarization of cyberspace by state actors, and (c) reluctance to advance binding agreements due to a lack of trust and mutual transparency regarding the real intentions of the parties. An example of such a vicious circle is evident in the context of international fora where initiatives to govern deleterious dynamics for international peace and stability from cyberspace are discussed. The presence of duplications within the United Nations, where the General Assembly has created the UN groups operating in the field of information and telecommunications in the context of international security—the Open-Ended Working Group (OEWG) and the UN Governmental Group of Experts (GGE)—to discuss the same issues, is a glaring example. To confirm this, consider that at the time this book is published, both UN groups, despite having issued vague and ambiguous proclamations recognizing the need to govern threats arising from the malicious use of ICTs, are essentially stuck on "how" and "which" part of existing international law to apply in cyberspace. These are example on how the International Community being trapped by the issue of anonymity and victims of polarization due to geopolitical disputes (ASPI 2022; Martino 2021).

A final noteworthy political element is given by the diffusion and distribution of power in the cyber context. Empirical evidence shows that the threshold of access to capabilities in cyberspace is low compared to other dimensions where violence is manifested. In particular, as Nye (2011) rightly pointed out, with the advent of cyberspace, there has been a diffusion of power unparalleled in any era before the digital one. While, in the past, there was a classic *translatio imperii* (power shift from one political entity to another), today we see the diffusion of power within the community of state actors and toward the outsider of the international arena. This situation are facilitating the action of non-state actors, including criminal groups and individuals (Nye 2011). The main implication of this subversion is not only in terms of "who" holds power, but also in terms of "how" conception of power is changing from that has traditionally been used in the context of international politics. In other words, with the advent of cyberspace, power it seems to be no longer measurable exclusively in terms of military resources, national GDP, human resources, or the quality of leadership. Instead, alongside such categories and units of power analysis, an independent variable is introduced. This variable is characterized by the widespread capacity (facilitated by cyber tools) to achieve objectives while maximizing benefits with the least possible cost, in economic, military, and human terms (Petit, 2020). If we want to use a doctrinal analogy, commonly employed in the context of strategic studies, although the Clausewitzian approach persists in traditional military domains - where the concentration of economic-military force is indispensable to achieve political objectives- the indirect approach theorized by Liddell Hart (1925) appears to hold greater relevance in the cyber context (Martino 2018; Van Puyvelde & Brantly 2019; Buchanan 2020; Whyte & Mazanec 2023). In this approach, the greatest success is achieved by targeting the adversary's "Achilles' heel," and in the context of the digital society, this is represented by the information dimension. This should not lead to the misconception that cyberspace has rendered the role of the state and other traditional domains (land, sea, air, and space) obsolete or that it will be the sole arena for intra- and international disputes or future warfare (Buchanan 2020; Whyte & Mazanec 2023; Martino 2023). On the contrary, awareness should lean toward recognizing that this domain, far from being considered an exclusive realm *only for technical experts*, requires an active and fundamental role by the state and its institutional apparatus to govern phenomenon and dynamics produced by cyberspace.

Statecraft in Cyberspace: The Case of Italy

In Italy, the topic of cybersecurity has been addressed, from an institutional and organizational perspective, for almost a decade starting from 2013. This year represents the issuance of the decree by the President of the Council of Ministers that established the first national architecture dedicated to cybersecurity. The adoption of this governmental decree has fostered the Italian development of appropriate national organizational cyber capabilities which, alongside classic tools such as military, diplomatic, cultural, and economic measures, have found a consolidated position in the pursuit of national interests (Tabansky and Ben Israel 2015). As illustrated in Fig. 1.1, to date, most states have equipped themselves with a national cybersecurity strategy to pursue and protect their interests in the context of cyberspace.

Similar to other nations, Italian political decision-makers have chosen (albeit, as we will later discuss, sometimes inconsistently) to establish an institutional framework and governance processes specifically aimed at managing the cyber phenomenon. Since 2013, this course of action has mostly been justified by the recognition of Italy's dual active role in the cyber domain. On the one hand, the role of ensuring national and domestic security, protecting citizens, social well-being, and safeguarding critical infrastructure responsible for the vital functions of democratic institutions from threats and risks originating from cyberspace. On the other hand, Italy (although only from the second decade of the 2000s) has also decided to complement the risk-oriented approach with the pursuit of opportunities arising from the digital society. In particular, Italian decision-makers have recognized the

Fig. 1.1 The diffusion of national cybersecurity strategies 2024. (Source: elaborated by the author, Luigi Martino 2024)

economic and social advantages of the ICT world, envisioning an investment plan, both economic and programmatic, in the field of research and technological development.

However, this dual strategy, focusing on threat prevention and seizing opportunities, largely arises from external pressures rather than choices driven by the maturity of the national political class (with a few rare exceptions that we will mention). It was driven by: (a) the necessity to keep pace with other countries considered "mature" in organizational bureaucracy, and (b) while simultaneously cultivating an ecosystem to position Italy significantly in technological development. As supported by an interviewe the main aim of this choice was "to overcome the inferiority complex resulting from short-sighted economic policy choices and the inability of the industrial and academic sectors to understand the potential that would have resulted from the information revolution" (Interview, Rome 2024). The paradigm shift was initiated in two distinct phases that led to an institutional and normative development, somewhat atypical compared to other European and international cases. Indeed, the decision to equip Italy, with a national architecture entirely dedicated to cybersecurity, arises from a need induced both by the European Union and NATO, namely, the priority to enhance the management, mitigation, and prevention capabilities of cyber risks in light of international events (primarily, but not only, the events in Estonia in 2007). As we will see in Chap. 3, it is worth highlighting that already from the early 2000s, Italy sought to manage cyber risks through a significant, and in some cases exclusive, role of homeland security (i.e. law enforcement) to protect critical information infrastructure from cyberattacks. As we have already mentioned, it is only since 2013 that Italy has decided to embark on the path of structuring a national cybersecurity governance, recognizing this matter among the prerogatives of national security and therefore the responsibility of the Prime Minister. This change in the allocation of responsibilities to the Prime Minister is an element that, as it will be illustrated later in Chap. 3, will have repercussions on the peculiar Italian choice to initially entrust the entire operational management of national cybersecurity to intelligence services and then decide, in the early 2020s, to build an ad hoc Italian National Cybersecurity Agency.

Structure of the Book

This book endeavors to critically analyze the excursus that Italy has undertaken in the context of cybersecurity from a governance and policymaking perspective implemented over the period 2013–2024. However, the historical analytical time-frame includes also policies implemented in 2005 (when the first ministerial decree on the protection of critical information infrastructure was issued). A preliminary caveat must be specified: the temporal analysis represents a dynamic snapshot as both the subject matter and the national (and international) context lend themselves to daily dynamism and are therefore impossible to fix in a precise static timeframe. This book serves a dual purpose: (a) contributing to filling an literature gap useful for reconstructing the regulatory, policy, and governance path implemented by Italy in the context of cybersecurity and (b) clarifying the present Italian national architecture and ecosystem dedicated to the governance of cybersecurity. Methodologically, the book combines primary and secondary sources, enriched by 35 semi-structured interviews with elite participants to empirically validate observations. The analysis begins with the premise that Italy's development in this field has been non-linear, facing numerous challenges primarily due to governmental instability and various external and internal influences. As we will see in the next pages, these influences are mostly derived from external factors, such as regulations and strategic decisions made by the EU and NATO, rather than from internal political awareness. The book is structured as follows: Chapter 2 reconstructs the Italian cyber ecosystem in order to provide a useful analysis to understand the various roles played by different actors in the Italian national system, from political, economic, and research perspectives. Chapters 3 and 4 will focus on a historical reconstruction of the policies implemented by Italy in the context of cybersecurity, aiming to understand the temporal trajectory that led to the establishment of the present national architecture—from where Italy started to where it stands today in this context. The third part is dedicated to the most innovative and debated legislation implemented by Italy in this context, namely, the establishment of the national cybersecurity perimeter. Chapter 5 seeks to illustrate the strengths and weaknesses of this legislation, beyond just the legal aspects, with a primary focus on the practical effects of this initiative. The final part analyzes the organizational and operational aspects of the National Cybersecurity Agency, inaugurated in 2021, which also led to the implementation of the National Cybersecurity Strategy. Here, it is necessary to specify the second caveat of this book: individual cyber security competencies attributed to central or peripheral bodies at the institutional level will not be analyzed. Instead, the book seeks to clarify the role played by the National

Cybersecurity Authority (i.e. ACN) in order to also understand the spill-over effects created by the policies implemented in Italy over the years. For these reasons, in Chap. 6, initiatives implemented by the Agency, as well as the limits and opportunities of the new national governance system in the field of cybersecurity, are examined along with the strategic objectives outlined in the national strategy. Finally, Chap. 7 will summarize the main outcomes of the book and the concluding remarks, with the attempt to foresee which possible scenarios may impact cyber security in Italy.

References and Additional Readings

ASPI. 2022. *Australian Strategic Policy Institute, UN Norms of Responsible State Behaviour in Cyberspace*. Guidance on Implementation for Member States of ASEAN. https://www.aspi. org.au/report/un-norms-responsible-state-behaviour-cyberspace.

Buchanan, Ben. 2020. *The Hacker and the State: Cyber Attacks and the New Normal of Geopolitics*. Cambridge, MA: Harvard University Press.

Bull, H. 1977. *The Anarchical Society*. London: Macmillan.

Carr, Madeline. 2015. Power Plays in Global Internet Governance. *Millennium: Journal of International Studies* 43 (2): 640–659.

———. 2017. *Cyberspace and International Order*, 162–178. https://doi.org/10.1093/ oso/9780198779605.003.0010.

Chemuturi, Murali. 2010. *Mastering Software Quality Assurance: Best Practices, Tools and Technique for Software Developers*, ix. J. Ross Publishing.

Choucri, Nazli. 2012. *Cyberpolitics in International Relations*. MIT Press.

Clarke, Richard A., and Robert K. Knake. 2010. *Cyber War: The Next Threat to National Security and What to Do About It*, 32. New York: Harper Collins.

D'Angelo, Gabriele, and Giampiero Giacomello. 2023. *Cybersicurezza. Che cos'è e come funziona*. Bologna: Il Mulino.

Dunn-Cavelty, Miriam. 2008. *Cyber-security and Threat Politics: US Efforts to Secure the Information Age*. New York: Routledge.

Eriksson, Johan, and Giacomello Giampiero. 2006. The Information Revolution, Security and International Relations: (IR)relevant Theory? *International Political Science Review* 27: 221–244.

Eriksson, Johan, and Giampiero Giacomello. 2023. Rise of the Nerd: Knowledge, Power and International Relations in a Digital World. In *Digital International Relations: Technology, Agency and Order*, 73–95. London, Routledge.

Giacomello, G., and B. Verbeek. 2023. Foreign Policy of Middle Powers. In *The Oxford Handbook of Foreign Policy Analysis*. Oxford: Oxford University Press.

Gray, Colin S. 2013. *Making Strategic Sense of Cyber Power: Why the Sky Is Not Falling*. Carlisle: US Army War College Press. https://press.armywarcollege.edu/monographs/529.

Kello, Lucas. 2013. The Meaning of the Cyber Revolution: Perils to Theory and Statecraft. *International Security* 38 (2): 7–40.

Kuehl, Daniel T. 2009. From Cyberspace to Cyberpower: Defining the Problem. In *Cyberpower and National Security*, ed. Franklin D. Kramer, Stuart Starr, and Larry K. Wentz. Washington, D.C.: National Defense University Press.

Libicki, Martin C. 2007. *Conquest in Cyberspace*. Cambridge, UK: Cambridge University Press.

Hart, Liddell. 1925. *Paris or the Future of War*. E.P. Dutton.

Maness, Ryan C., and Brandon Valeriano. 2016. The Impact of Cyber Conflict on International Interactions. *Armed Forces & Society* 42 (2): 301–323. https://www.jstor.org/stable/48670248.

Martino, Luigi. 2018a. La quinta dimensione della conflittualità. L'ascesa del cyberspazio e i suoi effetti sulla politica internazionale. In *Cyberspazio. La quinta dimensione delle interazioni umane*, 61–76. Società Editrice il Mulino.

———. 2018b. Cyber diplomacy e relazioni internazionali: le iniziative diplomatiche per mitigare il rischio di escalation militare nel cyberspazio. In *Il ruolo dell'Italia nella sicurezza cibernetica. Minacce, sfide e opportunità*, ed. Valerio De Luca, Giulio Terzi di Sant'Agata, and Francesca Voce.

———. 2019. *Confidence Building Measures (C.B.M.) in campo cyber: Attuali limiti e possibile contributo nazionale alla loro condivisione e applicazione*. Centro Alti Studi per la Difesa, Centro Militare di Studi Strategici.

———. 2021. *Le iniziative diplomatiche per il cyberspazio: punti di forza e di debolezza*. IAI Papers, no. 21/13. Istituto Affari Internazionali.

———. 2023. La guerra nel XXI secolo: la dimensione cyber e il conflitto russo-ucraino. In *La guerra tiepida: Il conflitto ucraino e il futuro dei rapporti tra Russia e Occidente*, ed. Andrea Manciulli and Enrico Casini. Koinè.

———. 2024. *Cyberspace Evolution and AI the International Competition and the Regional Posture in GCC*. LUISS University Press.

Martino, Luigi, De Zan, and Tommaso; Giacomello, Giampiero. 2021. Italy's Cyber Security Architecture and Critical Infrastructure. In *Routledge Companion to Global Cyber-Security Strategy*, ed. S.N. Romaniuk and M. Manjikian. Routledge. ISBN: 9780429399718.

Moore. 1993. *Assessing the Impacts of Changes in the Information Technology R&D Ecosystem: Retaining Leadership in an Increasingly Global Environment*, 2009. Washington, DC: The National Academies Press. National Academies of Sciences, Engineering, and Medicine.

Nye, Joseph. 2011. *The Future of Power in the 21st Century*. Public Affairs Press.

Perrow, C. 2011 [1984]. *Normal Accidents: Living with High Risk Technologies*. Princeton: Princeton University Press.

Petit, Nicolas. 2020. *Big Tech and the Digital Economy: The Moligopoly Scenario*. Oxford University Press.

Rid, Thomas. 2011. *Cyber War Will Not Take Place*. London: Hurst.

Tabansky, Lior, and Isaac Ben Israel. 2015. *Cybersecurity in Israel*. Springer.

Valeriano, Brandon, and Ryan C. Maness. 2014. The Dynamics of Cyber Conflict Between Rival Antagonists, 2001-2011. *Journal of Peace Research* 51 (3): 347–360.

Van Puyvelde, Damien, and Aaron F. Brantly. 2019. *Cybersecurity: Politics, Governance and Conflict in Cyberspace*. Cambridge and Medford, Polity.

Whyte, Christopher, and Brian Mazanec. 2023. *Understanding Cyber-Warfare Politics, Policy and Strategy*. Routledge.

Chapter 2
The Italian Cybersecurity Ecosystem

2.1 Introduction

Any complex organization, be it a state, a community, a biological organism, or a private enterprise, is based on an ecosystem, within which there are dynamics of direct and indirect interactions among various actors and the context in which this ecosystem is immersed (Moore 1993).[1] This concept holds true, and perhaps is even more crucial, in the context of cybersecurity, where it is not possible to conceive a one-directional approach or silos, where each actor moves according to their own needs independently of those of other actors. In particular, the concept of cybersecurity requires a series of broad and diversified interactions, within an ecosystem comprised of actors from the public, private, and civil society sectors, who, while bearing their own interests, share the common goal of contributing directly or indirectly to the cybersecurity of a country.

In light of this definition, to achieve a resilient posture in the cybersecurity field, the national ecosystem should be the prerogative of a series of initiatives capable of (a) involving and facilitating a country's actions in both internal and international contexts, (b) promoting multilateral and bilateral agreements, (c) incentivizing projects implemented by the private sector, (d) fostering training programs and skill development in ICT and cybersecurity, and (e) supporting research and development initiatives in cybersecurity (ITU 2021). All of these initiatives should be encompassed within a clear institutional and governance structures.

---

[1] Moore has defined the business ecosystem as *an economic community supported by a foundation of interacting organizations and individuals- the organisms of the business world. This economic community produces goods and services of value to customers, who are themselves members of the ecosystem.* See Moore (1993) and also National Academies of Sciences, Engineering, and Medicine. 2009. Assessing the Impacts of Changes in the Information Technology R&D Ecosystem: Retaining Leadership in an Increasingly Global Environment. Washington, DC: The National Academies Press. https://doi.org/10.17226/12174

This national ecosystem, in order to produce the desired effects, should be the result of clear objectives that a government aims to achieve through interaction with a heterogeneous multitude of stakeholders (governamental, non-governmental, civil society, individual, etc.), with the goal of ensuring an adequate sense of security among these actors regarding the dynamics arising from cyberspace. Starting precisely from national governance, an effective national cybersecurity ecosystem should be able to clearly identify the roles and responsibilities of the actors tasked with pursuing national interests in cyberspace (Tabansky and Ben Israel 2015; Cornish 2021), and these actors should be given the authority to execute strategic objectives (ITU 2021).

This latter aspect entails two implications: on the one hand, a clear governance framework based on rules and policies capable of creating a mechanism useful for identifying the government entities involved in implementing the pursuit of strategic objectives and, on the other hand, a distinct willingness on the part of the national political leadership to commit resources (funding, time, and personnel) to achieve the desired objectives (Cornish 2021).

We can thus deduce that the fundamental element of the national cybersecurity ecosystem lies, from one point of view, in intra-governmental collaboration (i.e. public-public partnerships), where institutional coordination and cooperation are fundamental functions, necessary to ensure that the combination of governance architecture and resources allocated for this purpose produce the desired results (ITU 2021). On the other hand, another fundamental pillar of the national cybersecurity ecosystem is represented by the role played by the private actors, the education and research sector, and civil society (i.e. multistakeholders dialogue). It follows that an effective (and efficient) national cybersecurity environment should envisage a clear principle of inclusion. This principle of inclusion is integral to the concept of public-private partnerships. When properly implemented, these partnerships can ensure a transparent system for verifying the dependencies between the government, the private sector, and other national non-governmental stakeholders, able to create a safer and more resilient ecosystem (D'Angelo and Giacomello 2023; ITU 2021; Cornish 2021).

However, establishing a sustainable ecosystem over time requires that the involved stakeholders have a clear understanding of the objectives and mutual benefits included in the context of the initiatives implemented. Based on the concept of "working together" (ITU 2021) both the public-public and public-private partnerships need to be able to influence the decision-making process in a linear and equal environment. Another strength of a national cybersecurity ecosystem lies in the set of initiatives implemented to ensure responsible and accessible innovation, initiatives based on the national environment's ability to facilitate the development or expansion of school and university curricula aimed at accelerating the development of cybersecurity skills through an interdisciplinary/multidisciplinary approach (D'Angelo and Giacomello 2023; Dunn Cavelty and Wenger 2022; ITU 2021; Abbott and Snidal 2000; Risse-Kappen et al. 1999; Finnemore and Sikkink 1998), thus ensuring a spillover effect on civil society (ITU 2021).

The ultimate goal of analyzing a national cybersecurity ecosystem should be to provide clarity in understanding the organization, governance, and processes that arise from the approval and implementation of national cybersecurity policies. This clarity enables to map the actors involved in the ecosystem and understand the existing legal and regulatory framework, including strategic objectives and operational aspects. Precisely, with the aim of trying to transpose in an empirical key "who does what," this chapter aims to explore in detail the actors composing the Italian cybersecurity ecosystem, identifying their role and interactions, as well as responsibilities and activities.

2.2   Defining the Italian Cybersecurity Ecosystem: Key Actors and Roles

The concept of a national cybersecurity ecosystem remains largely unexplored today. This further complicates the challenge of acquiring primary sources in the field of cybersecurity, a domain frequently enveloped in state or industrial secrecy, thus hampering its systematic study. However, in this chapter, we will attempt to reconstruct, to the extent allowed by the available sources, the roles of the actors and their respective responsibilities, as well as the initiatives that contribute to shaping Italy's cybersecurity ecosystem. Drawing from the term used by Moore (1993), we can define the Italian ecosystem as the presence of a community formed by various stakeholders within organizations and interacting individuals, constituting the entities of the national cybersecurity world. This community produces goods, services, rules, and policies deemed valuable for citizens, civil society, and institutions, all of which are integral members of the ecosystem. Specifically, this ecosystem focuses on the complex relationships among policymakers, researchers, educators (and their institutions), businesses (both large and small), cybersecurity clients (consumers, businesses, governments), and powerful contextual forces such as regulatory and legal frameworks, the supply of financial, human, and intellectual capital, economic infrastructure, and international competitive pressures in producing cybersecurity goods and services.

All these relationships contribute to creating economic wealth, social benefits, and a country's international posture in cyberspace. From a theoretical perspective (NAP 2009), an ecosystem can be deemed "healthy and vibrant" in the context of cybersecurity based on the following qualitative and quantitative indicators:

1. The quality and quantity of national maturity it generates over time.
2. The economic value it creates.
3. The level of security of goods and services.
4. The number and quality of jobs it creates.
5. Its resilience, understood as its ability to adapt to changing environmental conditions.

6. Its ability to create an internal collaborative environment and compete with other ecosystems at the international level.

It is important to note that the availability of precise measurements for these variables is not always guaranteed. However, a thorough analysis of each provides sufficient information to form a judgment on the relative health of Italy's national cybersecurity ecosystem, both in comparison to the past and to other countries, as well as in relation to its potentials or limitations. Another caveat of this research is that, like all ecosystems, the national cybersecurity one is complex and dynamic, involving many actors and multiple types of relationships, as depicted in Fig. 2.1.

In particular, as shown in Fig. 2.1, Italy's cybersecurity ecosystem is populated by various actors, ranging from individuals (e.g., students, researchers, civil society, etc.) to institutions (central and peripheral), public and private entities (providing essential services), private actors (constituting the national supply chain), and to the education and research sector (high schools and universities). Historically, the

Fig. 2.1 The Italian national cyber ecosystem. (Source: elaborated by the author, Luigi Martino 2024)

creation of the national cybersecurity ecosystem can be traced back to the period 2013–2023 when Italy began to establish its first national cybersecurity architecture (2013) and stabilize the new governance framework resulting from the establishment of the National Cybersecurity Agency (ACN) (2021) and the publication of the national cybersecurity strategy (2022). In general, we can define the Italian cybersecurity ecosystem based on consolidated interactions at various levels described in Fig. 2.1.

From an institutional and national governance perspective, within this ecosystem, the Prime Minister holds a position within the framework that serves as the main authority for policy direction and decision-making regarding cybersecurity, with the possibility of delegating responsibility to a designated authority. In his role, the Prime Minister oversees a group of key ministers in the context of national security and innovation, also involved in cybersecurity efforts (see Chap. 3). Moreover, additional specific actors are involved in the corresponding policy making and institutional tasks, such as:

- The Interministerial Committee for Cybersecurity (CIC), chaired by the Prime Minister and composed of government officials, plays a role in shaping cybersecurity policies and strategies at the national level. Through advisory functions, proposal development, and oversight, this committee aims to address cybersecurity challenges and ensure alignment with strategic objectives. It also has a hand in implementing cybersecurity strategies by providing rigorous supervision to ensure their success.
- The National Cybersecurity Agency (ACN) encompasses various roles: a coordination authority that oversees the coordination of actors, regulations, certifications, and sectoral oversight at a strategic level.
- The National Coordination Center (NCC) is responsible for coordinating cybersecurity efforts across technological and research sectors.
- The intelligence sector conducts intelligence activities in the field of cyber intelligence to safeguard military, economic, scientific, and industrial interests by analyzing cyber threats that may have a potential impact on national security.
- The Ministry of the Interior (MoI) serves as the national authority for internal security (homeland security) and leads initiatives to prevent and combat cybercrime through the Postal and Communications Police.
- The Ministry of Defense oversees defense and military security matters for the state by formulating policies and coordinating governance in cyberspace.
- The Ministry of Foreign Affairs and International Cooperation (MAECI) focuses on cyber diplomacy efforts to promote rights and freedoms in cyberspace.
- The Ministry of Economic Development supports industry transformation through the operations of Digital Innovation Hubs.
- The Ministry of Economy and Finance (MoEF), through the role of the Financial Police (i.e., *Guardia di Finanza*), works to combat financial crimes involving information technologies. The Bank of Italy issues rules and guidelines to improve the cyber resilience of the entities it supervises while also receiving incident reports from banks and financial institutions.

- The Ministry of Education and the Ministry of University and Research focus on promoting educational programs to address skills gaps in the labor market.
- The Department for Digital Transformation (DTD) coordinates government efforts to develop a strategy for the modernization of public administration.
- The Agency for Digital Italy (AgID) promotes innovation throughout the country and encourages the use of digital tools in government operations and interactions with citizens and businesses.
- The Cybersecurity Nucleus (NSC) ensures coordination among the aforementioned actors and develops action plans for cybersecurity. In the private sector, collaboration occurs between businesses, academic institutions, research organizations, and civil society through partnerships to improve the management of national ICT resources.

2.3   Evaluating the National Cyber Ecosystem: Governance Maturity and Structural Challenges

Italy's approach to building its cybersecurity posture has involved efforts spread over more than a decade, with the starting point dating back to 2013. Starting from this period, Italy began to develop the foundation of skills and institutional, economic, and research organizational capabilities in the field of cybersecurity, with a clear role assumed by the Prime Minister and various Italian governmental entities. This initiative entailed a system in partnership with private actors, building on an experience already initiated in 2005 in the context of cybercrime prevention and combating and in 2007 with a fundamental role entrusted to intelligence services in the context of national security protection (see Chap. 3). From these years onward, the Italian cybersecurity sector has grown rapidly, as demonstrated by the data in Graph 2.1.

In particular, according to official data (Cybersecurity & Data Protection Observatory 2023; Statista 2023), in the context of cybersecurity, there are over 1300 companies generating direct and indirect revenues of approximately 2 billion euros, with more than 40,000 qualified jobs.

Another fundamental element for understanding Italy's state of health in the economic and innovative context is related to the degree of digitalization penetration. Regarding connectivity, there has been progress both in the adoption of broadband services and in network distribution. However, there are still gaps in the coverage of Very High Capacity Networks, which is still below the EU average and the Digital Decade target for universal coverage by 2030. The majority of Italian small- and medium-sized enterprises (60%) have at least a basic level of digital intensity, and notably there has been significant growth in the use of cloud services (Neri et al. 2023; ISTAT 2023). In the context of digitalization, according to DESI (2023; 2024), Italy ranks 18th out of the 27 European Union countries as illustrated in Graph 2.2.

€ MLN *Graph 1: The Italian expenditure on cyber security years 2016- 2022*

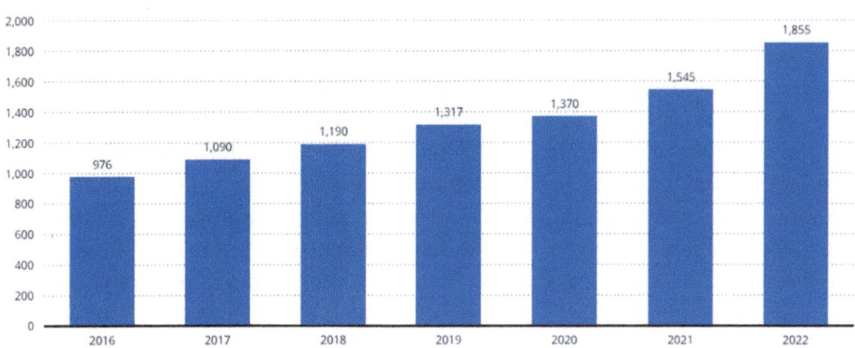

Graph 2.1 The Italian cybersecurity expenditure. (Source: elaborated by the author, Luigi Martino 2024))

	Italy		EU
	rank	score	score
DESI 2022	18	49.3	52.3

Digital Economy and Society Index (DESI) 2022 ranking

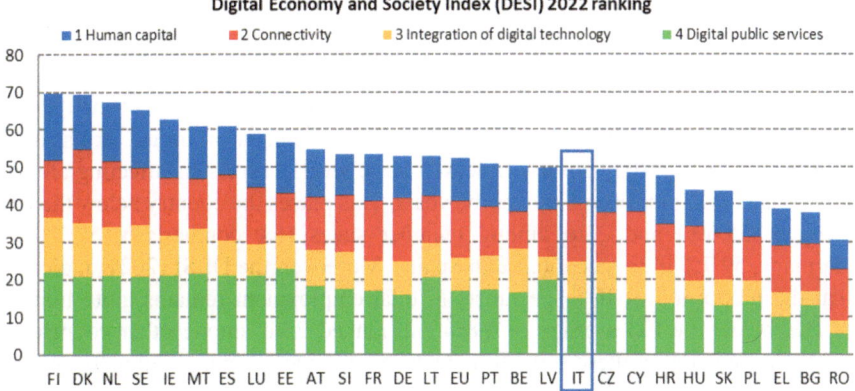

Graph 2.2 Digital Economy and Society Index 2022 Italy. (Source: DESI 2022)

Indeed, by cross-referencing data provided by DESI with those from other sources (Ambrosetti 2022; I-Com 2023; ISTAT 2023; DESI 2024), it emerges that Italy is experiencing a positive trend regarding national digitalization plans. This positive increase indicates a maturity achieved also in the context of awareness among Italian policymakers and leadership, partly attributable to the governance structure defined in 2021. Specifically, from a purely organizational perspective, digital issues in general, and cyber ones in particular, have gained political importance in recent years, especially following the establishment of a national authority dedicated to cyber resilience (i.e., ACN), the revision of the national cybersecurity strategy, and the initiation, in the wake of the National Recovery and Resilience Plan, of many project actions at both central and peripheral levels.

However, analyzing the data from DESI 2022–2023, a negative structural (almost social) element emerges that could compromise Italy's future capabilities in terms of competition and national preparedness in the context of the digital society. In fact, according to the DESI 2022–2023 survey, more than half of the Italian population lacks basic digital skills. These data place Italy in 25th position out of 27 EU countries in terms of human resources. The gap with the European average is less pronounced regarding people with advanced digital skills compared to those with basic skills (23% in Italy compared to 26% in the EU). Furthermore, Italy records a low rate of graduates in ICT: only 1.4% of Italian graduates dedicate themselves to ICT programs, which is the lowest among EU member states. In the labor market, the share of ICT professionals is 3.8% of total employment, remaining below the European average (4.5%). At the same time, only 15% of Italian companies offer ICT training to their employees, five percentage points below the EU average (I-Com 2023; DESI 2022). These data are confirmed also for the year 2023–2024. Indeed, according to data provided by DESI (2024) Italy is weak in data analysis, used by only 26.6% of companies compared to a European average of 33.2% (ranking 17th among EU member states). Even lower is the percentage of companies using artificial intelligence, at only 5%. However, it is important to note that this situation is significantly below the 2030 targets, even within the European average, where it reaches just over 8% (with respective targets of 60% for Italy and 75% for the EU). Additionally, the business sector highlights the difficult scalability of companies: in 2023, nationally, there are only 7 'unicorns,' less than 3% of the total in the EU. One of the country's major shortcomings remains in digital skills: only 45.8% of people in Italy possess at least a basic level, with issues affecting all age groups and little improvement over the years. This places the country below the EU average (55.6%) and 23rd among member states (for comparison, 52.2% in Germany, 59.7% in France, 66.2% in Spain). Moreover, the DESI (2024) indicates that the proportion of ICT specialists in total employment remains quite limited (4.1%, below the EU average of 4.8% and ranked 23rd among other countries), while the demand for these skills from businesses is increasing, pointing to a growing mismatch between demand and supply in the labor market. Definitely, the report stated that "the country's major gaps remain in digital skills, impacting efforts to close the digital divides and hindering competitiveness. Despite the roadmap's focus and Italy's numerous recent initiatives [...] ICT specialists in employment remains limited, while demand by enterprises for these skills is surging" (DESI 2024, 6). Moreover, the situation concerning investments in cybersecurity adds up to these data. In fact, despite cybersecurity remaining a priority investment in the digital sector in Italy, the country continues to rank last among G7 members, in terms of the ratio between the cybersecurity market and GDP (Cybersecurity & Data Protection Observatory 2023). These data are also confirmed by the low diffusion of cybersecurity insurance in Italy, as shown by the survey conducted by

Eurostat in 2022 (see Graph 2.3), which states that only 15% of Italian companies are insured against cyber incidents.

These data provided by Eurostat are also confirmed by the annual survey conducted by the Bank of Italy within the context of Italian businesses. According to the data provided by the Bank of Italy (2023), during the biennium 2021–2023, 22.8% of Italian companies did not allocate any budget for prevention activities against cyberattacks. Of these companies, two-fifths justify their decision by stating that they have not suffered any attacks, and one-sixth of those that have been violated by cyberattacks have not established any corporate function dedicated to the governance and management of cybersecurity and operational continuity. These companies, which largely coincide with those that spend less on prevention measures, are predominantly small in size and are mainly concentrated in Southern Italy (Bank of Italy 2024).

Another critical point of the national ecosystem concerns the lack of a consolidated strategy to support startups in the cybersecurity sector. According to government data (Cybersecurity & Data Protection Observatory 2023; MIMIT 2024; I-Com 2024), the few Italian startups operating in this sector during the biennium 2020–2022 raised an average of about $1 million, a significantly lower figure compared to the European average of just over $3 million and the global average of around $18 million. Another index of a country's innovative capacity is also given by the ability to foster innovation through an environment capable of stimulating the creation, growth, and operational continuity of startups (Tabansky and Ben Israel 2015).

In this regard, Italy has not yet managed to build a mature environment for the valorization of startups capable of integrating research and development of

Enterprises having insurance against ICT security incidents, 2022
(% enterprises)

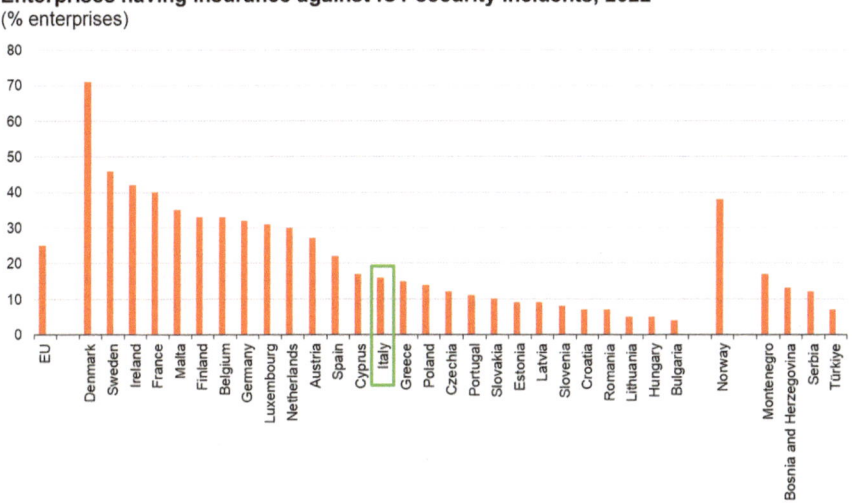

Graph 2.3 Level of Italian cyber insurance expenditures. (Source: Eurostat 2022)

innovative solutions even in the context of cybersecurity. For example, in 2023, the number of innovative startups[2] amounted to 14,029,

with a decrease of 233 units (−1.6%) compared to the previous quarter (MIMIT 2024). When it comes to new companies operating in the cybersecurity sector, a survey conducted by I-Com in 2023 identified 121 startups across the country. Looking at the year of foundation of these companies, there is a steady growth until 2019, followed by a significant slowdown in 2020 coinciding with the COVID-19 pandemic. However, the following year saw a significant recovery with 35 startups, the highest number of cybersecurity startups founded in a single year. But, in 2022, there was an equally significant contraction, with only 11 new startups in this field, a trend that continued in 2023 with only four new innovative startups founded in that year. Regarding the geographical distribution of these startups, the North of the country prevails overall with its 56 startups, followed by the Center with its 43, and then the South with only 22 (I-Com 2024). In light of the data collected, a critical structural element of the Italian ecosystem emerges in failing to foster pioneering entrepreneurship in a context like cybersecurity, where research and innovation are the cornerstone of these activities. In other words, although institutional governance has reached a certain level of maturity, this advantage risks being nullified by political, economic, and industrial shortcomings, unable to promote a fertile ground for the creation of an innovative environment, dedicated to the development of solutions useful for pursuing the country's technological independence, and ensuring better resilience and national security. To enhance the economic drivers resulting from opportunities in cybersecurity, in 2023 the ACN launched a strategic initiative called the Cyber Innovation Network (CIN). This program is designed to support innovative businesses and maximize the utilization of research findings. In its initial phase, the agency issued a public notice to identify qualified entities, such as startup incubators or accelerators, that have specialized expertise in cybersecurity, with the aim of facilitating the commercialization of research and innovation. To enhance the potential economic drivers derived from opportunities in cybersecurity, the ACN initiated a strategic initiative in 2023, named the Cyber Innovation Network (CIN).

---

[2]According to Decree-Law 179/2012, the recognition of innovative startup status is reserved for capital companies established for less than 5 years, with an annual production value of less than five million euros, unlisted, and meeting specific criteria for technological innovation established by national regulations (summary of requirements and benefits). Among the approximately 384,000 capital companies established in Italy in the last quarter of 2023 and still operational, 3.7% were registered as innovative startups at the time of the survey. The financial independence index of innovative startups is 0.39, lower than that recorded by other non-innovative new enterprises (0.44). If only innovative startups and profit-making capital companies are considered, a value of 0.37 is observed against 0.44. On average, for every euro of production, innovative startups generate 21 cents of value added, a figure lower than other companies (27 cents). However, limiting to profit-making companies, startups generate higher value added compared to capital companies: 33 cents versus 29.

2.4   The Role of Research and Efforts Toward a National Excellence Hub

Research, innovation, and education are three critical components essential for ensuring the sustainability of a national ecosystem from both a human resources perspective and in terms of spillover effects in the economy. Notably, the DESI 2024 report identifies a low level of skills as a major issue for Italy within the ICT sector, which includes cybersecurity. To address this, Italy has incorporated research as a cornerstone of its National Cybersecurity Strategy. Specifically, two of the 82 strategic measures are aimed at promoting research with industrial repercussions. Measure #53 aims to enhance Italy's industrial and technological autonomy concerning strategically significant computer products and processes, safeguarding national interests in the sector. This includes the development of proprietary algorithms and the advancement of new national cryptographic capabilities. Measure #54 seeks to support research and development, particularly in new technologies, by promoting the incorporation of cybersecurity principles. It supports this through funding, both public and private investments, and simplification mechanisms, encouraging cybersecurity projects by the private sector—especially focusing on startups and innovative SMEs—as well as by Competence and Research Centers operating nationally. These two measures underpin the Research and Innovation Agenda for Cyber Security 2023–2026, which was jointly approved by the ACN and the Ministry of University and Research (MUR) in 2023. This Agenda outlines six areas of intervention: Area #1: Data security and privacy Area #2: Management of cyber threats Area #3: Security of software and platforms Area #4: Security of digital infrastructures Area #5: Societal aspects Area #6: Governance aspects. Universities, research centers, spin-offs, and other actors involved in the development of cybersecurity technologies collaborate with public institutions and private companies to promote the development of national capabilities. The main aim of this trajectory included in the above mentioned strategic measures is to enable Italy to take a leading role as an attractive and competitive player in cybersecurity. These initiatives, sometimes independent and implemented sporadically, are also accompanied by some that can be defined as best practices at the European and international levels.

For example, considering training as a fundamental element to ensure quality research and a pool of skills, according to surveys conducted by the Institute for Competitiveness (I-Com 2023), Italy has initiated a series of initiatives aimed at enhancing the offers in cybersecurity in higher education and university programs. During the period covered by this book, from January 2023 to March 2024, a total of 250 university courses focused on cybersecurity or the implications of cybersecurity at a multidisciplinary level were identified. This data shows "an increase compared to the 91 courses reported at the beginning of 2022" (I-Com 2023). These courses include lectures within bachelor's and master's degree programs, as well as specialized programs such as master's and doctoral degrees (CINI 2024). Among the 97 public and private universities recognized by the MUR, it was observed that there are 270 units offering courses related to cybersecurity for the academic years 2022/2023 and 2023/2024 (MUR 2024). However, the growing number of courses on cybersecurity does not automatically guarantee a high level of specialization or in-depth study of the subject matter. This gap often arises from cybersecurity education typically being incorporated into broader degree programs, particularly at the master's or postgraduate level, while the provision is significantly lower in bachelor's and master's degree programs, indicating a lack of strategy by Italian universities in ensuring adequate preparation over 5 years of academic training. Another issue in the present Italian cybersecurity educational offer is the fact that out of 57 university courses focusing on this sector, their distribution highlights "a greater presence in regions such as Lazio and Piedmont, with 29 universities not offering any specific courses on cybersecurity" (I-Com 2023).

In addition to these geographical discrepancies, there is another issue entirely related to university strategies: according to the available data for 2023–2024, the distribution of courses among departments reveals that the majority of the offering is structured in engineering departments, followed much less prominently by sectors of the social sciences (political science, law, economics) or other sciences (mathematics, chemistry, and architecture).[3]

Regarding applied research activities in the context of cybersecurity, a strong impetus to such activity has come from the exceptional funding provided by the National Recovery and Resilience Plan (PNRR) financed by the European Union in response to the severe crisis resulting from COVID-19. In this context, Italy has proposed and developed various projects in different sectors (from medicine to AI, robotics, environmental sustainability, etc.). Specifically in the field of cybersecurity, the case of the extended partnership "SERICS—Security and Rights in the CyberSpace," funded with 116 million euros from European funds, emerges. The SERICS Foundation's projects are divided into ten "spokes," each focusing on specific cybersecurity themes such as (a) human, social, and legal aspects; (b) disinformation; (c) attacks and defenses; (d) security of operating systems and virtualization; (e) encryption and security of distributed systems; (f) security of software and platforms; (g) infrastructure security; (h) risk management and governance; and (i) digital transformation and data protection.

---

[3] There are exceptions, such as the case of the ICT Policies and Cyber Security course held from 2017 to 2023 at the Cesare Alfieri School of Political Sciences of the Department of Political and Social Sciences at the University of Florence. Starting from the 2016–2017 cohort, a virtuous model based on the adaptation of the course was initiated, allowing access to students from the faculties of computer engineering and statistics, who were enabled to participate in the program alongside colleagues from international relations, European studies, and communication strategies. An experiment was initiated by the author of this book that, over the course of 7 years, led to the multidisciplinary training of over 500 students and resulted in approximately 90% of those who had attended, passed, and supported the exam finding employment in both technical and humanistic cyber security fields. Another interesting exception: it is represented by the second cycle degree/2-year master in digital innovation policies and governance implemented by the Department of Political and Social Sciences at the University of Bologna.

As described in Table 2.1, each project is led by its own leader, with universities at the forefront, along with the contribution of various affiliates, both public and private.

Another relevant case in the Italian context of research and applied training within national cybersecurity is the CINI (National Interuniversity Consortium for Informatics), created in 1989 under the supervision of the Ministry of Education, University and Research, which represents a consortium formed by Italian universities involving more than 1300 professors from 45 public universities.

The CINI is a *sui generis* entity in the Italian context because, through its framework, it manages to support joint research activities with universities, higher education institutions, research centers, industries, and public entities, facilitating access and participation in research and scientific development projects as well as technology transfer. Within the framework of CINI, the National Cybersecurity Laboratory stands out, with 53 of the main Italian universities and research institutes, including more than 500 professors and researchers dealing with both computer science and aspects related to policies and international relations, as the case of the Lab-Node Center for Cyber Security and International Relations Studies (CCSIRS).[4] One of the specific programs to address the lack of skills is represented by both the CyberChallenge initiative and ITASEC, the Italian conference entirely dedicated to cybersecurity. CyberChallenge is an initiative aimed at identifying, attracting, recruiting, and placing professionals in the field of cybersecurity. Between 2017 and 2023, there has been a significant increase in participation requests from students in schools and universities, going from a total of 683 registrations to 4720 (CINI 2024). Consequently, there has been an expansion of venues and institutions involved, as well as a significant increase in the admission process to the program.[5] Furthermore, the SERICS Foundation and CINI organize two free courses open to young people, called OliCyber and CyberTrials, aimed at consolidating digital skills and encouraging young people to approach the field of cybersecurity.[6]

---

[4] The author of this book is the Director of the Center for Cyber Security and International Relations Studies. Founded in 2015 at the University of Florence, the CCSIRS is a node of the National Cybersecurity Laboratory of CINI, in addition to the Computer Science node of the University of Florence. The CCSIRS deals not only with research in the field of cyber diplomacy and dynamics of cyberspace in international politics, but also with research projects such as CyberReadiness, which aims to measure human risk in organizations, and the CISO Executive Program project in collaboration with the Dubai Cyber Innovation Park, a unique case at the GCC regional level and one of the few cases at international level recognized also as a best practice by the World Economic Forum.

[5] The program consists of 12 weeks of training, managed independently by each location, primarily focused on aspects such as cryptography and security concerning hardware, networks, software, and the web.

[6] According to the organizers, OliCyber represents the first national cybersecurity competition aimed at high school students, offering them the opportunity to participate in an exclusive educational program covering all the key skills needed to address digital threats. On the other hand, CyberTrials is a school aimed at all female students of high secondary schools, including those who do not have previous computer knowledge. Participants are guided by high-level experts along a path that covers both technical and nontechnical aspects of the digital world, from threat identification to major forensic investigation techniques and ethical implications in the computer context (CINI 2023).

Table 2.1 Projects and actors involved in *SERICS—Security and Rights in the CyberSpace*

Spoke	Research aims and objectives	Leader	Associates
1. Human, social, and legal aspects	The detailed objectives of AT 1 can be classified into five macro-categories. The first category concerns rights, rules, definitions, taxonomies, and authorities useful for creating new forms of co-regulation in cyberspace. The second category analyzes the legal and ethical issues of cybersecurity, such as fundamental rights within the new ecosystem. The third category includes continuous learning and educational models on legal issues in cybersecurity. The fourth category analyzes cybercrime and cyber diplomacy as important and crucial elements of a new national strategy, aiming to develop knowledge on this topic in the public, not just academic, sphere. The fifth category considers digital sovereignty, also in relation to computations and technologies based on artificial intelligence, cloud, fog, and edge computing and their applications in specific sectors such as energy and transportation	CNR	UNICA, UNI GE, UNIBO, UNIMI, Sant'Anna Pisa, UNIFI
2. Disinformation and fake news	The objective is to implement methodologies for analyzing textual and multimedia content to develop models for identifying disinformation attempts. Additionally, the analysis of social media communities will highlight the cognitive vulnerabilities of participants and threats related to the spread of fake news. The goal is to design an early warning system to alert against false information, leveraging the syntactic integrity of content and models related to disinformation flows. The resulting framework aims to raise awareness about the risky effects of sharing questionable content. Moreover, the framework will support experts and security professionals in decision-making, adopting a human-in-the-loop approach	UNISA	CNR, CNIT, UNICA, UNIVE, UNIMI, UNI ROMA 1, ENI S.p.A., Scuola IMT Alti Studi Lucca

(continued)

Table 2.1 (continued)

Spoke	Research aims and objectives	Leader	Associates
3. Attacks and defenses	The detailed objectives can be divided into four macro-categories: (i) development of advanced tools for malware analysis and software aimed at identifying vulnerabilities that could be exploited by malware themselves; (ii) development of tools for network traffic analysis to identify communications related to ongoing attacks; (iii) development of machine learning systems resilient to attacks and through which it is possible to extract knowledge aimed at creating more advanced tools for timely analysis and early detection of attacks; and (iv) analysis of "human factors" involved in an attack with the development of tools for analysis and correlation of information from OSINT (open-source intelligence) and for defense and prevention of attacks based on social engineering techniques	UNICA	SSSUP, CNR, UNIBA, UNI GE, UNICAL, UNI ROMA 1, UNISA, UNIVE, ENI S.p.A., Leonardo S.p.A., Telsy S.p.A.
4. Operating system and virtualization security	It deals with developing high-level automatic security services as well as innovative methodologies for security evaluation and assurance to support secure-by-design development and verification of cloud, edge, and 5G applications. The effectiveness of the proposed techniques will be evaluated through stress tests in simulated but highly realistic attack scenarios, securely performed within a federated CyberRanges platform	UNIGE	CINI, CNIT, CNR, Scuola IMT Alti Studi Lucca, Fondazione Bruno Kessler, Fondazione Ugo Bordoni, UNISA, UNICAL, UNI ROMA 1, Fincantieri, Leonardo S.p.A.
5. Cryptography and security of distributed systems	Research activities are focalized in the domains of cryptography and security of distributed systems. Various subtopics are considered: (i) cryptographic primitives and protocols; (ii) foundational cryptography and cryptanalysis; (iii) post-quantum cryptography; (iv) digital identity, authentication, and accountability; and (v) distributed ledgers and blockchain	UNICAL	Fondazione Bruno Kessler, UNISA, CNR, UNICA, PoliTO, Deloitte, Intesa Sanpaolo S.p.A

(continued)

Table 2.1 (continued)

Spoke	Research aims and objectives	Leader	Associates
6. Software and platform security	The objective is to develop new formal techniques based on secure compilation and secure composition to bridge the gap between formal models, essential for providing full correctness guarantees and real-world implementations. The second scientific objective is to provide innovative solutions to protect the software supply chain, including software management and development processes. The aim is to develop new techniques for conducting security tests through continuous dynamic analysis and for protecting software by detecting harmful activities and preventing or limiting their impact, following a self-defensive paradigm	UNIVE	UNISA, UNICA, Deloitte, UNIBA, UNIFI, UNI ROMA 1, Scuola Imt Alti Studi Lucca
7. Infrastructure security	It holds four specific objectives: (i) design and develop an open, secure national computer architecture, which will be the starting point for building secure infrastructures that do not suffer from potential risks arising from the use of proprietary technologies; (ii) improve the security of automotive infrastructure, which, with the massive interconnection and electrification of cars, will become one of the country's most vulnerable assets; (iii) improve the security, protection, and resilience of smart power grids, which are fundamental for optimizing energy consumption and achieving the Green Deal; and (iv) contribute to improving the security posture of ICT assets (e.g., networks, IT/OT systems and services) included in the "National Cybersecurity Perimeter," by providing ontologies, methodologies, guidelines, best practices, and appropriate tools	PoliTO	CINI, CNR, Fondazione Ugo Bordoni, Scuola IMT Alti Studi Lucca, Scuola Superiore Sant'Anna Di Pisa, UNICA, UNI GE, Deloitte, Leonardo S.p.A., Telsy S.p.A.
8. Risk management and governance	The objective is to address both scientific-technological and legal and political challenges through new models for continuous assessment of threats and vulnerabilities but also through the design of self-defensive network components. AT 8 also aims to promote the vision that an evolved digital Europe requires the protection of fundamental rights and freedoms, the promotion of social awareness and widespread computer education, and achieving gender balance in cybersecurity	UNIBO	CNIT, CNR, UNIBA, UNIFI, UNI GE, UNITO, UNIMI, Deloitte

Table 2.1 (continued)

Spoke	Research aims and objectives	Leader	Associates
9. Securing digital transformation	The objective is to promote research on (i) the development of decentralized finance solutions based on secure distributed technologies such as distributed ledgers and smart contracts; (ii) strengthening the security properties of data and privacy in services provided by public administration within e-government programs; (iii) proposing remote health solutions based on personal devices, essential for more efficient management of chronically ill patients or those requiring continuous monitoring; and (iv) developing quantum key distribution technologies for critical applications	UNI ROMA 1 "La Sapienza"	CNR, UNIBA, UNICA, UNI GE, UNIMI, UNISA, Telsy S.p.A., Intesa Sanpaolo S.p.A
10. Data governance and protection	The objective is to develop an ecosystem for sharing, availability, and control of data, ensuring their selective and secure sharing while ensuring functionality, efficiency, scalability, and sovereignty	UNIMI	UNI ROMA 1, UNIFI, UNICA, UNISA, Leonardo S.p.A.

Source: Elaborated by the Author with the information provided by the Foundation "SEcurity and RIghts In the CyberSpace" (SERICS 2024)

In addition to these initiatives, another economic driver of the Italian cyber ecosystem is represented by the "Digital Europe Programme" aimed at facilitating the digitalization and the cybersecurity awareness in the SME and Public Administration through the role of the European Digital Innovation Hubs—EDIHs. The aim has been to facilitate the digital transition through the adoption of advanced digital technologies such as artificial intelligence, high-performance computing, and cybersecurity.

2.5   Conclusions

The dynamics of interactions between an ecosystem made up of various actors and stakeholders and an institutional governance architecture represent a crucial barometer for assessing the success or failure of the Italian approach to cybersecurity. In particular, the activities implemented collectively and individually in over the past 10 years have paved the way for an Italian cyber ecosystem that, like other highly developed countries, is following political-strategic choices which, in some cases, determine a contradictory path characterized by institutional maturity and social and economic-industrial criticalities. The impact of this dichotomy is even more relevant when considering that the field of cybersecurity, immersed in the context of technological evolution, has undergone (and will continue to undergo) profound transformations, influencing every aspect of our lives, including the maintenance of commercial activities, education, employment, healthcare, production, government operations, national security, communications, entertainment, science, engineering, and even warfare.

Despite the inherent complexity due to the multitude of actors involved in the national cybersecurity landscape, the challenge is to keep the ecosystem well tuned in order to produce the expected benefits and define a virtuous circle, where distortions due to complexity are resolved non-traumatically in cyclical phases. For example, risks arising from regulatory dynamics can create dangerous anomalies for the entire ecosystem, because some companies may incur high compliance costs due to regulatory burdens and are thus called upon to make investments that not all private entities can afford (such as SMEs or startups). In this way, there exists not only an intrinsic vulnerability in the stability of the national ecosystem due to a top-down regulatory approach but also the creation of market competition distortion because of substantial disparities between those who can afford the costs and those who are excluded from the market. These are just some examples that, as we will see in the sections dedicated to national regulations, can be remedied not only through the creation of clear governance, but also through the implementation of an effective public-public and public-private partnership system among the various components of the national cybersecurity ecosystem.

References and Additional Readings

Abbott, Kenneth W., and Duncan Snidal. 2000. Hard and Soft Law in International Governance. *International Organization* 54 (3): 421–456. http://www.jstor.org/stable/2601340.

Ambrosetti. 2022. The European House Ambrosetti. Rapporto annuale dell'Osservatorio. Online resource.   https://www.astrid-online.it/static/upload/28c8/28c84e025fdf5a10e545f8394b be4c37.pdf

Bank of Italy. 2024. L. Bencivelli and M. Mongardini, Italian firms' cybersecurity – Risk perception and mitigation strategies; Rome.

CINI. 2024. Consorzio Interuniversitario Nazionale per l'Informatica (Italian National Inter-University Consortium for Informatics). URL: https://www.consorzio-cini.it/index.php/en/about-us-eng/union.

Cornish, Paul, ed. 2021. *The Oxford Handbook of Cyber Security*. Oxford University Press.

D'Angelo, Gabriele, and Giampiero Giacomello. 2023. *Cybersicurezza. Che cos'è e come funziona*. Bologna: Il Mulino.

Digital Economy and Society Index (DESI). 2022. *The Digital Economy and Society Index (DESI) 2022*. Online resource. https://ec.europa.eu/newsroom/dae/redirection/document/88764

DESI. 2022. European Union Commission, The Digital Economy and Society Index (DESI), 2022. URL: https://digitalstrategy.ec.europa.eu/en/policies/desi.

DESI 2024 Report on the state of the Digital Decade 2024; URL https://digitalstrategy. ec.europa. eu/en/library/report-state-digital-decade-2024.

Dunn Cavelty, Myriam, and Andreas Wenger, eds. 2022. Cyber Security Politics: Socio-Technological Transformations and Political Fragmentation.

Eurostat. 2022. Eurostat, Enterprises having insurance against ICT security incidents, 2022, (online data code: isoccisce ic).

Finnemore, Martha, and Kathryn Sikkink. 1998. International Norm Dynamics and Political Change. *International Organization* 52 (4): 887–917. http://www.jstor.org/stable/2601361.

I-Com. 2023. Rapporto Osservatorio sulla Cibersicurezza: L'ecosistema italiano della sicurezza informatica tra regolazione, competitività e consapevolezza—Politiche, competenze, regole. Istituto Per la Competitività. https://www.i-com.it/wp-content/uploads/2023/02/Rapporto_Osservatorio-sulla-Cibersicurezza_2023.pdf

———. 2024. Report on the state of the Digital Decade 2024; URL: https://digitalstrategy. ec.europa.eu/en/library/report-state-digital-decade-2024.

International Telecommunication Union (ITU). 2021. Guide to Developing a National Cybersecurity Strategy: Strategic Engagement in Cybersecurity, 2nd Edition. https://www. un.org/counterterrorism/sites/www.un.org.counterterrorism/files/2021-ncs-guide.pdf

ISTAT Istituto Italiano di Statistica. 2023. Imprese e ICT | Anno 2022. Online resource. https://www.istat.it/it/files//2023/01/REPORTICTNELLEIMPRESE_2022.pdf

MUR. 2024. *Ministry of University and Research*. Source online in Italian https://www.mur. gov.it/it/areetematiche/universita/lofferta-formativa-titoli-rilasciati/accreditamento-lauree-e-lauree.

MIMIT. 2024. Cruscotto di Indicatori Statistici - Startup innovative Report con dati strutturali 4° trimestre 2023, Elaborazione 01.gennaio.2024 URL: https://www.mimit.gov.it/images/stories/documenti/4_trimestre_2023.pdf.

Moore. 1993. *Assessing the Impacts of Changes in the Information Technology R&D Ecosystem: Retaining Leadership in an Increasingly Global Environment*, 2009. Washington, DC: The National Academies Press. National Academies of Sciences, Engineering, and Medicine.

NAP. 2009. National Research Council, "Ecosystems, Ecosystem Services, and Biodiversity". In *Advancing the Science of Climate Change*. Washington, DC: The National Academies Press. https://doi.org/10.17226/12782.

Neri, Martina, Federico Niccolini, and Luigi Martino. 2023. Organizational Cybersecurity Readiness in the ICT Sector: A Quanti-Qualitative Assessment. *Information and Computer Security* 32 (1).

Osservatorio Cybersecurity & Data Protection. 2023. Lo Scenario della Cybersecurity in Italia nel 2023: Report Cybersecurity & Data Protection. https://www.osservatori.net/it/prodotti/formato/report/report-scenario-cybersecurity-italia-2023

Risse-Kappen, Thomas, Steve C. Ropp, and Kathryn Sikkink. 1999. The power of human rights: International norms and domestic change. *Foreign Affairs* 78: 143.

Serics. 2024; Online sources accessible at https://serics.eu/progetti/.

Statista. 2023. Market Insights: Cyber Security in Italy. Data based on 2023.

Tabansky, Lior, and Isaac Ben Israel. 2015. *Cybersecurity in Israel*. Cham: Springer.

Chapter 3
Foundations of Cybersecurity Policy and Governance in Italy

3.1 Introduction

This chapter critically analyzes the historical trajectory of the Italian cybersecurity governance and policymaking foundations. The content of the chapter offering insights into the Italian regulatory and governance evolution in the field of cybersecurity, critical information infrastructure protection (CIIP) and attribution of competencies among different governmental and institutional bodies. The analysis spans a period marked by the issuance of the first ministerial decree protecting information critical infrastructure in 2005 up to the reform of intelligence services in 2007. It addresses the challenge of capturing the fluidity of cybersecurity within a static and bureaucratic timeline. The chapter begins by outlining the early stages of Italy's cybersecurity initiatives, which were significantly influenced by international variables: events such as terrorist attacks in US (9/11) and Europe (Madrid 2004 and London 2005) and decision made by actors such as NATO and subsequent EU regulations. It discusses the pivotal reforms and policy shifts that have shaped Italy's national cybersecurity architecture, highlighting the move from a fragmented approach focused on CIIP critical infrastructure protection under law enforcement, to a more centralized intelligence-driven national security strategy. This transition is examined through various legislative actions and strategic decisions that, progressively, integrated cybersecurity into national security concerns, under the direct responsibility of the Prime Minister. This chapter serves a dual purpose: (a) contributing to filling a literature gap useful for reconstructing the regulatory, policy, and governance path implemented by Italy in the context of cybersecurity and (b) clarifying how the present Italian national architecture dedicated to the governance of cybersecurity has been consolidated. The analysis proposed in the chapter starts from the assumption that Italy's journey on an institutionalized cybersecurity has not been linear, but rather has been featured by a series of obstacles, mainly due to governmental instability. Another observation is related to the fact that the main inputs primarily derived from external factors (such as regulations and strategic choices of both the EU and NATO), rather than internal political awareness.

3.2 Origins of Italian Cybersecurity Policies

The implementation journey of Italy's institutional framework dedicated to cybersecurity governance has been marked by fluctuating phases, sometimes exhibiting clear dichotomies, largely influenced by endogenous and exogenous factors. Among the endogenous elements that have negatively impacted the linear development of a mature and comprehensive cybersecurity governance, it is worth mentioning the political instability of Italian governments, the dissolution of parties, and the absence of systematic public investments in the cybersecurity context.

Governmental instability is certainly not unique to Italy and, in some respects, it can be seen as an added value of democratic dynamics (Veenendaal 2022). However, in the Italian context, it is notable that from 1946, the year of the foundation of the Italian Republic, to 2022, there have been 68 governments and 31 Prime Ministers. This instability has had a negative impact on a series of capabilities for programming, implementing, and consolidating policies that have affected the real maturity and efficiency of national cybersecurity governance. In particular, according to some observers (Interviews Rome and Milan, 2023–2024), the fundamental problem stems from subjugating the competence of national cybersecurity to the political figure that must rely on parliamentary stability, namely, the Prime Minister (officially the President of the Council of Ministers). In particular, this political prerogative allows the Prime Minister to appoint or dismiss the operational figures responsible for cybersecurity management, making them subject to the spoils system. Therefore, as it will emerge in this chapter, instability is entirely at odds with what is required for the management of cybersecurity at the strategic and operational levels, which includes, among other things, a medium- to long-term period of political commitment and administrative/leadership continuity (De Zan, Giacomello, Martino 2021).

At the same time, Italian political awareness has been marked by a chronic lack of investments, or allocation of resources, dedicated to the implementation of policies approved in the context of cybersecurity. This is proved by the fact that the only budget allocated in the 4-year period 2013–2017 was the one-off budget allocated in 2016 (Draft Budgetary Plan 2016), amounting to approximately 150 million euros, while the two reforms (that followed in the 2013–2017 period) explicitly envisaged "zero-cost expenditure for their implementation" (De Zan, Giacomello, Martino 2021). Additional confirmation arises from the shift in direction coinciding with the allocation of funds from the European Union's Recovery and Resilience Plan, which was initiated in response to the COVID-19 pandemic.

The exogenous elements, that have positively influenced the development of an institutional cybersecurity governance, derive from Italy's participation in international or supranational organizations such as the North Atlantic Treaty Organization

(NATO) and the European Union (EU). Particularly, as we will see in the following pages, it is through its participation in these two Organizations that Italian policy-makers will find the impetus to adopt a national architecture entirely dedicated to managing dynamics arising from cyberspace. The contribution of endogenous and exogenous elements, as well as the awareness of the Italian political class, regarding the need to strengthen national cybersecurity, has allowed the transition from a het-erogeneous governance in the diffusion (and, in some respects, confusion) of responsibilities to a recent attempt to centralize functions through the creation of an Agency (i.e. ACN) entirely dedicated to national cybersecurity (see Chap. 6).

3.3 The Attribution of Responsibilities Between Law Enforcement and Intelligence Services

From a legislative perspective, we can trace the first Italian normative act in the cybersecurity field back to 2005 when, following the terrorist attacks in Europe between 2004 and 2005 (in Madrid and London), the Italian Parliament converted the decree of the then Minister of the Interior Pisanu into Law No. 155 titled "Urgent measures to counter international terrorism" (Law 155 2005). In this law, at art. 7 bis, it is established that the protection of critical information infrastructure (CIIP) is as follows:

> Without prejudice to the competencies of the information and security services, referred to in Articles 4 and 6 of Law No. 801 of October 24, 1977, the body of the Ministry of the Interior for the security and regularity of telecommunication services ensures the cyberse-curity services of critical computerized infrastructures of national interest identified by decree of the Minister of the Interior, operating through telematic connections defined by specific agreements with the responsible parties of the structures concerned. (art. 7 bis (Law 155 2005), Original translated by the author)

The choice made in 2005 reveals two limits in the Italian political decision-making process. On the one hand, the matter of cybersecurity was confined solely to the CIIP, presupposing that cyber risks only affect this type of infrastructure. On the other hand, a primary role is attributed to the Ministry of Interior, assuming that cyber activities are only of a criminal nature, without considering the political, eco-nomic, and military aspects. In other words, the 2005 choice immediately demon-strated both strategic and operational limitations, viewing the contrast of malicious activities in cyberspace as an evolution of classical crime and therefore the respon-sibility of law enforcement. The macroscopic limit resulting from a law enforcement-oriented approach, within the Italian regulatory context, is due to the need to act according to the principle of the "mandatory nature of criminal action" contained in Article 112 of the Italian Constitution. This principle has played a limiting role for private operators who "for fear of being brought to court or investigated for intent or negligence have decided not to share any cyberattacks or incidents" (Interview, Milan 2023).

However, despite these limitations, the legislative input of 2005 instigated important operational initiatives in the Italian ecosystem. A noteworthy initiative is the establishment, initially as a specialized unit and later formalized in 2008, of the National Anti-Crime Computer Center for the Protection of Critical Infrastructures (CNAIPIC), which operates within the Postal and Communication Police Service and is dedicated to the prevention and repression of cybercrimes directed against national critical infrastructures (CNAIPIC 2024).

3.4 The Paradigm Shift: From Homeland Security to an Integral Part of the National Security System

The international events resulting from the fall of the Berlin Wall in 1989, the terrorist attacks of September 11, 2001, in the United States, and those of 2004–2005 in Europe, as well as the advent of the first global-scale cyberattacks in Estonia, have had a substantial impact on the international politics of the twenty-first century. Triggered by these events that "expanded" the international arena to threats from non-state actors, in 2007 (during the same period that Estonia has been affected by the first massive cyber attacks against a state), Italy approved the reform of its intelligence services which, as highlighted by Soi (2014), found themselves tasked not only with political-military interests but also, and above all, with safeguarding Italy's economic, scientific, and industrial interests. With this awareness, Law 124/2007 drastically reforms the structure and functions of Italian intelligence services, by assigning the political responsibility for the national security sector solely to the Prime Minister, abolishing the sharing of responsibilities with the Ministries of Interior and Defense. The President, in line with the reform, can delegate tasks to an Undersecretary, or Minister, who will have a direct role in the context of activities falling within the scope of national security.

What emerges from the 2007 reform is the role of the Department of Information for Security (DIS), identified as the central body in the exercise of intelligence competencies. The reform created the Interministerial Committee for the Security of the Republic (CISR) for consultation with the Prime Minister. The Agencies, AISE and AISI, have divided competencies based on territorial threats, with the former handling external threats and the latter the internal ones. The reform also established new rules on state secrecy, balancing security and citizens' rights and providing functional guarantees to the "insiders" of the intelligence sector. The Parliamentary Committee for the Security of the Republic (COPASIR) oversees the security information system to ensure compliance with the Constitution and laws, with extensive powers and reporting obligations to the President of the Council of Ministers. The 2007 reform, thus restructures the institutional and governance framework of the Italian intelligence services, and particularly envisions a systemic governance related to the Italian Intelligence System for the Security of the Republic (Law 124/2007).

The new governance of the Intelligence System emerged from the 2007 reform includes:

- Prime Minister.
- Undersecretary to the Prime Minister for Intelligence (if appointed).
- CISR—Interministerial Committee for the Security of the Republic.
- DIS—Department of Information for Security.
- AISE—External Intelligence and Security Agency.
- AISI—Internal Intelligence and Security Agency.

As shown in Fig. 3.1 (representing the result of the architecture depicted in the 2007 reform) the new governance structure of Italian intelligence grants the Prime Minister a prerogative on national security. The Prime Minister can appoint an undersecretary with a delegation to intelligence and is supported by the CISR from a strategic-political perspective. Operationally, the DIS plays a functional and centralizing role, coordinating AISI and AISE for their respective tasks.

The reform of the Italian intelligence system has played a fundamental role in the context of cybersecurity, both from a political and operational responsibility standpoint. Indeed, while, as seen in the previous section, the 2005 legislation (the Pisanu law) recognized a primary role for law enforcement functions regarding the defense of critical information infrastructure (CII), the 2007 Law No. 124 represents a paradigm shift. This shift can be simplified as follows: considering the Prime

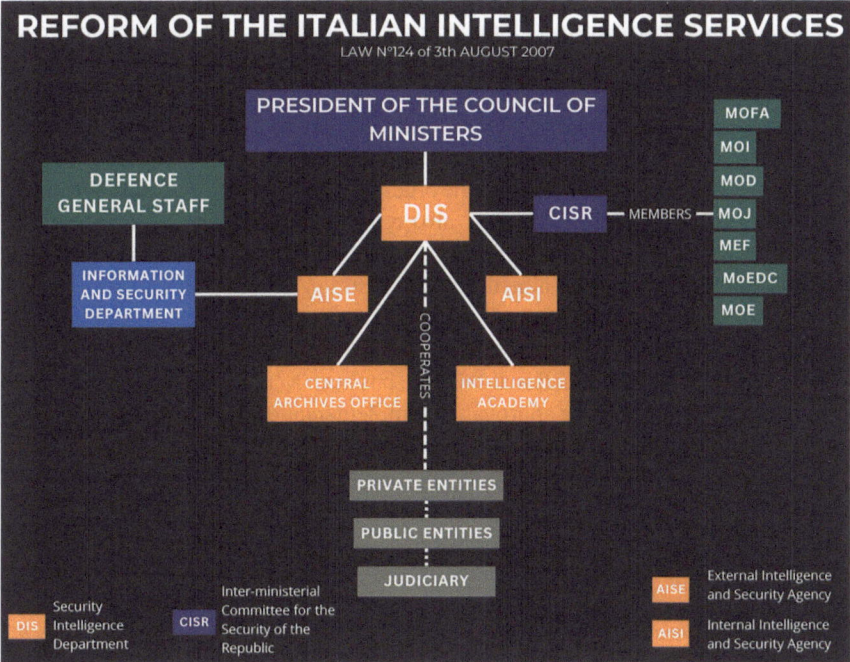

Fig. 3.1 The reform of the Italian secret services and its governance. (Source: elaborated by the author, Luigi Martino 2024)

Minister's direct responsibility for national security issues and recognizing that the cyber issue falls within the scope of these prerogatives, Intelligence services are therefore assigned the function of protecting national interests in cyberspace. In this way, cybersecurity is encompassed within the broader concept of national security and the protection of the country's system. This prerogative is explicitly stated in Law No. 124/2007, where it is specified that the Information Security Department (i.e. DIS) coordinates information research activities aimed at strengthening cyber protection and national cybersecurity. However, this paradigm shift from a law enforcement-oriented approach, to an approach based on the prerogative of national security, was crystallized only from 2012 onward. With the approval of Law No. 133 on August 7, 2012 (which amends and integrates Law No. 124/2007), it is specified that "it is the prerogative of the President of the Council of Ministers, having consulted the Interministerial Committee for the Security of the Republic (CISR), to adopt specific directives to strengthen information activities for the protection of critical material and immaterial infrastructures" (Law 133/2012), with particular regard to national cyber protection and cybersecurity.

3.5   Conclusions

Throughout this chapter, it has been possible to examine the "embryonic" initiatives implemented by Italy to define the allocation of responsibilities at the institutional level regarding cybersecurity. A journey initiated as early as 2005, which, however, from the outset has been influenced by internal political turbulence, financial constraints, and external factors. Indeed, the frequent changes of government in Italy have had a significant impact on cybersecurity policies, resulting in inconsistency and discontinuity of strategic vision, favoring a more reactive, rather than proactive, approach to cybersecurity. However, the crucial reform of 2007, which modified the organization of intelligence services, also represented a change by incorporating cybersecurity dynamics within national security, under the direct supervision of the Prime Minister. This change aimed to address strategic policy shortcomings while seeking to achieve a more centralized and coordinated governance system. This paradigm shift, from a law enforcement perspective to an interpretation of cybersecurity as an integral part of national security, has been a crucial element that allowed the initiation of a systematic approval and implementation process of policies, with the result to create the first national cybersecurity architecture.

References and Additional Readings

CINI. 2018. The Future of Cybersecurity in Italy: Strategic Focus Areas. https://www.consorzio-cini.it/images/Libro-Bianco-2018-en.pdf.
De Zan, Tommaso, Giampiero Giacomello, and Luigi Martino. 2021. Italy's cyber security architecture and critical infrastructure (Chapter 10). In *Routledge Companion to Global Cyber-Security Strategy*, 1st ed. Routledge. https://doi.org/10.4324/9780429399718-10.

Draft Budgetary Plan. 2016. Italy's Draft Budgetary Plan 2016. Available in English URL: https://www.mef.gov.it/export/sites/MEF/inevidenza/documenti/DRAFT_BUDGETARY_PLAN_2016-_EN_-22-10-2015-Completo_11_11.pdf.

Law. 124/2007. Only in Italian, Legge 3 agosto 2007 Sistema di informazione per la sicurezza della Repubblica e nuova disciplina del segreto. GU n.187 del 13-08-2007. URL: https://www.gazzettaufficiale.it/eli/gu/2007/08/13/187/sg/pdf.

Law. 133/2012. Modifiche alla legge 3 agosto 2007, n. 124, concernente il Sistema di informazione per la sicurezza della Repubblica e la disciplina del segreto. Only in Italian https://www.normattiva.it/uri-res/N2Ls?urn:nir:stato:legge:2012;133.

Law. 155 2005. Available only in Italian, Legge 31 luglio 2005, n. 155. Conversione in legge, con modificazioni, del decreto-legge 27 luglio 2005, n. 144, recante misure urgenti per il contrasto del terrorismo internazionale. Gazzetta Ufficiale della Repubblica Italiana.

Chapter 4
Evolution of Italy's National Cybersecurity Governance

4.1 Introduction

Between 2013 and 2021, Italy underwent significant reforms in its cybersecurity, which were pivotal in shaping the country's approach to a dedicated architectural governance. These changes were influenced by key events, including the adoption of a cybersecurity strategy by the European Union in February 2013, NATO's concentrated efforts to address political-military risks stemming from cyberspace, and a growing awareness among Italian policymakers regarding the critical importance of cybersecurity.

Initially, Italy concentrated on establishing a cybersecurity architecture to fortify its defenses against escalating threats. This architecture, put in place in 2013, underwent evaluations in 2015 to gauge its effectiveness and pinpoint vulnerabilities, leading to a reform in 2017, that underscored the vital role of intelligence services in cybersecurity (De Zan 2016). Furthermore, an examination of the strengths and weaknesses of Italy's cybersecurity approach, across the two architectures established between 2013 and 2017, yielded insights that influenced subsequent enhancements. These insights contributed to the establishment in 2021 of the National Cybersecurity Agency, charged with overseeing Italy's cybersecurity resilience. The chapter explores the operational and governance adjustments across the three national architectures implemented in 2013, 2027 and 2021, detailing the specific roles and responsibilities assigned to various governmental bodies. It also critically analyzes the impact of these reforms on the overall efficiency and effectiveness of the national cybersecurity policy, identifying both strengths and weaknesses in each architecture.

The institutional and governance overview delves into how and when Italy has bolstered its cybersecurity capabilities through policies and regulatory framework, adjustments and strategic adaptations over time.

4.2   The First National Cybersecurity Architecture

On January 24, 2013, the so-called "DPCM Monti" (DPCM Monti 2013) containing guidelines for cyber protection and national cybersecurity was approved. This governmental decree has institutionalized the first national architecture entirely dedicated to cybersecurity. Therefore, 2013 represents a turning point through which the first national cybersecurity architecture is established and as described in Fig. 4.1, the governance is organized into three levels of action: political-strategic, operational-functional, and tactical-reactive.

It is worth mentioning that, despite its bureaucratic complexity, the national architecture incorporates the regulatory actions implemented between 2007 and 2012 regarding Intelligence services, placing the cybersecurity domain within the context of national security (See Chap. 3). This choice, according to experts "is justified by the evidence that the cyber threats have now assumed a strategic dimension, so much so that these threats are considered by major international actors a risk factor of primary importance, directly proportional to the level of development reached by information technologies" (Interviews, Rome 2023 and Milan 2024).

As detailed in Fig. 4.1, at the top of the decision-making structure of the national architecture we find the President of the Council of Ministers (i.e. Prime Minister) assisted by the Ministers forming the Committee for the Security of the Republic

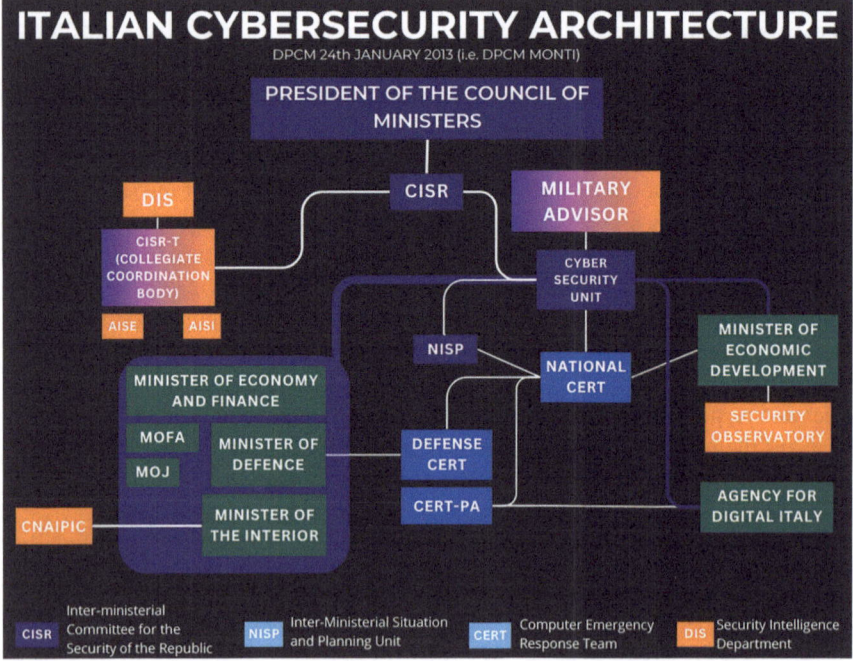

Fig. 4.1 The first Italian cybersecurity architecture and governance. (Source: elaborated by the author, Luigi Martino (2024))

(CISR or Political CISR). These key actors are invested with crucial roles: adopting the strategic decisions; defining the national cybersecurity strategy; and structuring a document dedicated to the "national strategic framework for the security of cyberspace" (National Cyber Security Strategy 2014). Moreover, they are responsible for issuing directives resulting from the implementation of the architectural governance (DPCM Monti 2013).

Alongside the Interministerial Committee, a collegial coordination body operates, headed by the director-general of the Department of Information for Security (DIS), playing a complementary role to the CISR and aptly termed "technical CISR." This body consists of leaders responsible for cybersecurity within the administrations represented in the CISR. During sessions related to cybersecurity, the military advisor to the Prime Minister joins this body. In the 2013 institutional architecture, the technical CISR assumes significant responsibility: it plays both an investigative role and a supportive role to the Political CISR during control phases to ensure the implementation of the "national plan for the security of cyberspace" (National Cyber Plan 2014). Additionally, it coordinates the institutional interactions among different administrations, competent offices, and public and private entities involved in the national cybersecurity ecosystem.

Within the institutional framework outlined by the "Monti Decree", the Intelligence bodies are called to play a particularly important role. These bodies, including the two Agencies (AISI and AISE), have a crucial role in the collection and processing of information. On the other hand, the DIS is primarily involved in formulating strategic analyses, evaluating and forecasting threats from cyberspace, and promoting a cybersecurity culture. Furthermore, regarding aspects related to prevention and preparation in case of crisis situations, the decree also establishes the Cybersecurity Unit (or Cybersecurity Cell NSC). The NSC, as established by the "Monti Decree", operates permanently at the Office of the Military Advisor and is chaired by the latter. The Unit is composed of representatives from the DIS, AISE, AISI, the Ministry of the Interior, the Ministry of Foreign Affairs, the Ministry of Defense, the Ministry of Economic Development, the Ministry of Economy, the Department of Civil Protection, and the Agency for Digital Italy.

In the 2013 architecture, the NSC primarily functions as a link between various institutional actors operating in the field of cybersecurity (Vestito 2018). Its main responsibility is to strengthen prevention, alert, and preparedness activities in case of crisis situations, and it has its own permanently active operational unit. Additionally, it is tasked with taking response and recovery actions in case of crises, and it can, if necessary, activate the Interministerial Cyber Crisis Table. In fact, the latter is an updated version of the preexisting Interministerial Situation and Planning Unit (NISP), established by a Decree of the President of the Council of Ministers in May 2010 for the management of national crises. Therefore, the NISP, as the Interministerial Table, is responsible for verifying and coordinating the management and response to cyber crises by the involved administrations. Finally, it is noteworthy the central role recognized for private operators in the institutional process dedicated to national security and crisis management in response to threats stemming from cyberspace. In particular, the "Monti Decree" echoes the indications of the European Union policy framework, involving operators who own or manage critical infrastructures of national and European relevance, whose criticality is closely linked to the integrity of ICT systems, and whose malfunctions can dire national security and citizens safety. In light of this initiative, for the first time a mandatory reporting system for any significant security or integrity violation of the IT systems of such infrastructures has been implemented, with directly involvement of the NSC or, if required, of the Intelligence services.

4.3   Evaluating the Initial National Cybersecurity Framework

From an organizational perspective, what stands out from the analysis of Fig.4.1 is a widespread operational-functional level involving a multitude of actors, coordinated partly by the DIS and partly by the Military Advisor to the Prime Minister. Both of these actors become, de facto, responsible for the prevention and management of cyber crises with the participation of other entities entitled to take part in the tactical-reactive level. In particular, the "Monti Decree" establishes, for the first time, an institutional architecture dedicated to Italian cybersecurity governance. However, strategic, operational, and tactical gaps are immediately apparent. Specifically, from a strategic-political perspective, a lack of unified political responsibility emerges due to shared decision-making between the Prime Minister and the CISR. At the same time, the operational level is diluted by a widespread attribution of responsibilities among various, more or less, relevant departments, leading to an overlap of competencies and roles, creating a lack of clarity at the tactical-reactive level. Additionally, despite the fundamental role of private entities being emphasized multiple times, including in the strategic objectives outlined in the National Strategy and the Implementation Plan published in 2014, the role of such actors is not explicitly foreseen in the national architecture. Another crucial point, also emphasized during interviews, is that "the legislation does not allocate any budget for the implementation and operational activities related to the purposes outlined by the national architecture" (Interviews, Rome 2024).

Italy has therefore decided to address the significant challenges in the definition and implementation of cybersecurity policies through an approach that is almost a political necessity, rather than the efficient implementation of substantive policies. Indeed, a key critical element in the governance of cybersecurity created in Italy in 2013 is represented by ambiguity in the decision-making process. While prominent roles at the political-strategic level have been well defined, the process of assigning operational competencies and responsibilities remains immersed in a nebulous path. This lack of clarity has led to situations of uncertainty, undermining the effectiveness of operational actions. The ambiguity in attributing competencies has also been favored by bureaucratic complexity, and the resulting slow decision-making, both of which have contributed to creating another significant criticality in the architecture implemented in the framework of the "Monti Decree". In fact, it immediately became clear that there was a huge gap between the objectives of the policy and the actual effectiveness and efficiency of the national architecture. This gap was favored by the choice to implement an operational level in which cybersecurity measures required the involvement of multiple entities, each of which had to follow complex procedures and respond to its own prerogatives. The decision-making slowness resulting from the intricate bureaucratic process of 2013, as we will see in the following pages, necessitated a revision of decision-making processes, in order to increase their capacity for risk management, mitigation, and prevention (De Zan 2018b).

A third critical element is the lack of coherence in cybersecurity strategy. While clear objectives have been defined at the political-strategic level, the implementation of these objectives, at the operational level, has lacked coherence. This dissonance has led to a fragmented application of cybersecurity policies, also resulting in the vagueness of implemented measures. Therefore, we can argue that, the main criticality arising from the 2013 architecture was the failure to ensure precise alignment between political strategy and operational actions. Another weakness that emerged from the implementation of the 2013 governance was operational complexity, which constituted one of the main practical shortcomings. The lack of a centralized operational direction and the deficit in information exchange, which is also due to the involvement of both the Military Advisor and the Intelligence services, have led to operational inefficiencies. These inefficiencies are particularly evident toward private operators who have not had direct access to the operational decision-making process or to preventive information. A macroscopic discrepancy between strategic objectives and action plans has also caused clear difficulty in pursuing the strategic objective of promoting an effective public-private partnership at the operational level. The biggest obstacle was precisely the confusion regarding decision-making hierarchies and vagueness regarding responsibilities. All these complexities at the operational decision-making level had the potential to engender vulnerabilities and risks to national security. The lack of a timely response to cyber threats could have jeopardized the country's security, necessitating a thorough review of the national cybersecurity governance.

4.4   The Second National Cybersecurity Architecture and the Enhanced Role of Intelligence

The organizational limitations, effectiveness issues, and efficiency constraints are evident in the 2013 architecture, which is based on a chaotic bureaucratic decision-making process and suffers from the absence of a budget. For these reasons, in 2015, a directive from the President of the Council of Ministers (the so-called Renzi Directive) was issued. Its aim was to assess the limitations of the 2013 DPCM and to improve the efficiency of the decision-making process and the effectiveness of actions entrusted by national governance (Renzi 2015).

The strength of the Renzi Directive was to uniquely identify the shortcomings of the architecture issued by the "Monti Decree", especially in light of initiatives undertaken at both the European and NATO levels. In the former case, the Italian architecture highlighted its limitations compared to the regulatory requirements of the Network Information Security Directive (NIS Directive) approved by the European Union in 2016 (NIS 2016). The Italian macroscopic problem became immediately clear when there was a need to identify a single national contact point. This point, as required by the NIS Directive, should coordinate the identification of operators of essential services, act as a collector of information and notifications in the event of cyber incidents at the national level, and interact with European counterparts (especially ENISA and the European Commission) to ensure effective information exchange. However, this objective clashed with the chaotic situation resulting from the 2013 governance. At the same time, following the 2016 Warsaw Summit, NATO (2016) openly declared that "cyber attacks constitute a clear challenge to the security of the Alliance and could be as damaging to modern societies as a conventional attack." For these reasons, cyberspace was recognized as a domain of military operations. This implies that all Members of the Atlantic Alliance are called upon to ensure NATO's ability to protect and "conduct operations in all these domains and to maintain our freedom of action and decision in all circumstances" (NATO 2016). In light of this, NATO requires its allies to have "more effective organization of cyber defense and better management of resources, expertise, and capabilities" (NATO 2016). Therefore, following the Renzi Directive, the DPCM dated February 17, 2017, was approved. Its aim was to reform both the governance structure and the distribution of decision-making power in the field of cybersecurity. As detailed in Fig. 4.2, the "Gentiloni DPCM" in theory consolidates a distributed and shared governance of cybersecurity (DPCM Gentiloni).

At the strategic apex, in continuity with the previous architetture, also in the DPCM Gentiloni the political decision-making level is placed with the exclusive responsibility of the President of the Council of Ministers, who always acts in collaboration with the CISR. Instead, the operational level is reformed to improve the decision-making chain, making it more responsive in terms of speed and readiness, to prevent and respond in case of cyber threats. Consequently, in the quest to identify a public administration body directly under the Prime Minister's office, capable of combining decision-making readiness with responsibilities in national security matters, the choice fell on the intelligence services, specifically the Department of Security Information (DIS).

The transfer of responsibility for the Cybersecurity Unit from the Military Advisor to the Intelligence services, represents a significant change in Italy's approach to cybersecurity. At the same time, the DIS is identified as the national point of contact for the European Union in the field of cybersecurity, within the provisions of the NIS Directive. The DIS is also designated as the coordinator for the implementation of community-derived regulations. This shift in institutional organization underlines the growing importance of cybersecurity, and the recognition that the cyber threat has become one of the main challenges to national security. The active involvement of the Intelligence services also indicates the maturity of

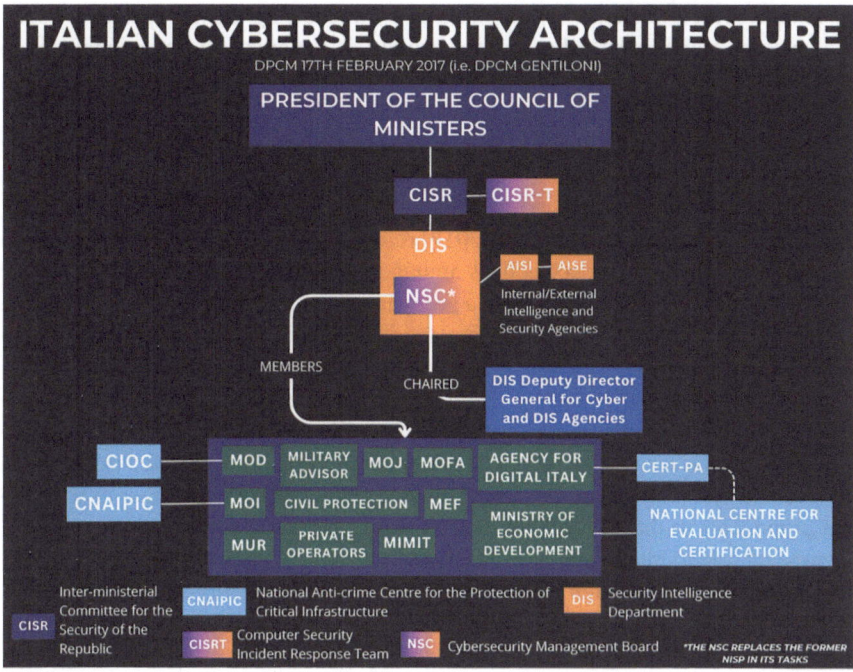

Fig. 4.2 The second Italian cybersecurity architecture and its governance. (Source: Elaborated by the author, Luigi Martino (2024))

Italian policymakers in considering cybersecurity a matter of national security. At the same time, the appointment of a deputy Director of the DIS for the first time with the responsibility to coordinate the Cybersecurity Unit is another important step to ensure the full mature cybersecurity governance.

4.5   Analysis of the Strengths and Weaknesses of the Second National Cybersecurity Architecture

Although the changes, introduced by the 2017 reform, bring evident improvements compared to the effects of the policies implemented in 2013, the architecture resulting from the "Gentiloni DPCM" has clear shortcomings. Among the most significant limitations are the overlap of responsibilities and the absence of a dedicated budget for the implementation, management, and innovation of operations carried out by the national governance. Another noteworthy element is the lack of an update to the National Cybersecurity Strategy, which, unlike the Operational Plan updated in 2017, remains unchanged despite the national and international changes that have occurred in the 4 years separating the two measures.

Another relevant element is emphasized by a recurring theme repeated by various experts, namely, the debate on the validity of assigning an operational role to Intelligence services, in a context that involves continuous information exchange rather than *the jealous and vital secrecy of information*. In other words, despite recognizing the validity of the choice to designate Intelligence as the operational arm of the national cybersecurity architecture - considered as an improvement to the "chain of command" compared to the 2013 governance - the structure resulting from the implementation of the 2017 reform has highlighted a polarization between two conflicting perspectives among the participants of the national cyber ecosystem. On the one hand, there is a group in favor of considering Intelligence the natural institutional body tasked with maintaining an operational role in Italian cybersecurity. This choice is justified by comparing Italy and other national architectures, implemented in highly mature countries from a digital, economic, and political perspective. The main assumption was that these countries have chosen to delegate cybersecurity activities to Intelligence services, as seen, for example, in France and Germany (Interviews, Rome and Milan 2024). On the other hand, there are proponents of the opposing viewpoint. who consider it a flaw in Italy to give a significant role to Intelligence in a context where it is necessary to implement information exchange that often cannot be subject to confidentiality needs (Interviews, Rome and Milan 2024). Critics of the choice contained in the "Gentiloni DPCM" especially highlight the operational dichotomy of the DIS regarding the need to balance the natural inclination of Intelligence services to keep information secret, and the requirements of operating in the cybersecurity ecosystem where information exchange is a vital element (Interviews Rome and Milan 2024). In particular, while the openness and sharing of information between Intelligence services and other governmental Agencies (such as law enforcement) can enhance collaboration and response to cyber threats, however "it is important to ensure that there is adequate control and monitoring to prevent abuse or misuse of information" (Interview, Rome 2023). Moreover, the fact that "the transfer of responsibility to Intelligence services could raise issues related to privacy and surveillance, as these services have traditionally operated in a context of secrecy," resulting in "interaction between a public administration operating in the context of state secrecy and other public and private actors requiring continuous dialogue can lead to a short circuit in the ability to provide effective responses" (Interviews, Rome 2023 and 2024).

This debate has been exacerbated between 2018 and 2019, especially in light of the approval of both European regulations (the domestication of the NIS Directive in the Italian legal system and the approval of the EU Cybersecurity Act) and national regulations (such as Law 133/2019, known as the national cybersecurity perimeter see Chap. 5). As we will see in the following pages, these regulations have imposed significant obligations on both public and private operators of essential services and critical infrastructure. These initiatives, in addition to resulting in an exponential increase in operational and functional obligations, have also granted an inspection and sanctioning role to intelligence services, tasked with ensuring the compliance with the regulations.

4.6   The Third National Cybersecurity Architecture and the Establishment of the National Cybersecurity Agency

2021 can be considered a milestone year for the organization and governance of cybersecurity in Italy. In fact, in light of the dynamics triggered by the COVID-19 pandemic, geopolitical tensions, and initiatives undertaken at the European level, Italian policymakers decided to renew and reformulate the national architecture. The ultimate goal of this new provision, as stated in the preamble of Decree-Law No. 82 of 14 June 2021, is to "address cyber challenges effectively and coordinately," and for this purpose, "it became necessary to redefine the national cybersecurity architecture" (Law 82/2021). Precisely to ensure an adequate level of coordination and centralization of supervisory capacities, the same Law also establishes the creation of the National Cybersecurity Agency, indicated as the national authority in the field of cybersecurity (see Chap. 6). Among the novelties introduced by the 2021 legislation are the definition of concepts that also assume legal value, such as the definition of "cybersecurity," which refers to "the set of activities necessary to protect networks, information systems, IT services, and electronic communications from cyber threats, ensuring their availability, confidentiality, and integrity and ensuring resilience, also for the purposes of national security and national interest in cyberspace" (Law 82/2021). Similarly, the definition of "National Resilience in Cyberspace" has been included, which refers to "activities aimed at preventing harm to national security, namely harm to the independence, integrity, or security of the Republic and democratic institutions established by the Constitution as its foundation, or to the political, military, economic, scientific, and industrial interests of Italy, resulting from the interruption or compromise of an essential function of the State or an essential service" (Law 82/2021). From a governance perspective, Fig. 4.3 summarizes the new national cybersecurity architecture resulting from Law 82/2021.

As illustrated in Fig. 4.3, at the top of the governance system remains the Prime Minister, who is entrusted with the overall direction and responsibility for "cybersecurity policies" and is responsible for adopting the relevant national strategic goals and appointing the leadership of the National Cybersecurity Agency. The Prime Minister may delegate the functions related to the cybersecurity (except those exclusively reserved for him/her) to a so called Delegated Authority of the Security Information System of the Republic. At the Prime Minister's office, the Interministerial Committee for Cybersecurity (CIC) is established, a body with advisory, proposal, and supervisory functions on cybersecurity policies. The CIC committee is composed as follows:

- The Prime Minister (who chairs it)
- The delegated authority, where established
- The Minister of Foreign Affairs and International Cooperation
- The Minister of Interior
- The Minister of Justice
- The Minister of Defense
- The Minister of Economy and Finance

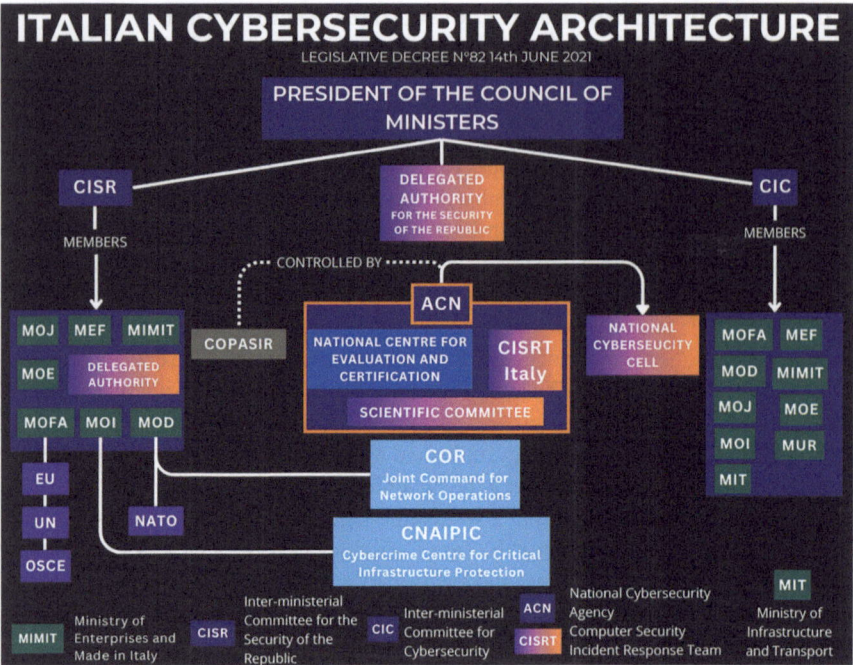

Fig. 4.3 The present Italian cybersecurity governance architecture. (Source: elaborated by author, Luigi Martino (2024))

- The Minister of Economic Development
- The Minister of Ecological Transition
- The Minister of University and Research
- The Minister delegated for technological innovation and digital transition
- The Minister of Infrastructure and Sustainable Mobility

It is interesting to note that the National Cybersecurity Agency, as established by the law, is legally constituted as a public law entity and has regulatory, administrative, asset, organizational, accounting, and financial autonomy.[1] Furthermore, according to the new architecture and governance system, the National Cybersecurity

---

[1] The National Cybersecurity Agency will serve as the single national contact point for public and private entities for security measures and inspection activities related to the security of networks, information systems, and electronic communication networks. Among other things, the National Cybersecurity Agency is in charge to structure the national cybersecurity strategy, it is the national authority for the certification of cybersecurity under Article 58 of Regulation (EU) 2019/881, and it is responsible for the qualification of cloud services for public administration. More generally, as we will see in the following pages, the National Cybersecurity Agency is vested with all cybersecurity-related powers previously attributed to the Ministry of Economic Development, the Prime Minister's Office, the Department for Information Security, and the Italian Digital Agency.

Agency promotes the development of national capacity to prevent, monitor, and mitigate cyber incidents and attacks, with the aim of enhancing the security of the ICT systems of entities included in the national cybersecurity perimeter (see next chapter).

In summary, the governance system of the national cybersecurity system, as established by Decree-Law No. 82 of 2021, emphasizes the concept of cyber resilience. This concept is essential to promote the development of the country's digitalization, productive system, and public administrations. Furthermore, resilience is considered essential to achieve autonomy both nationally and at the European level in the field of products and digital processes that are strategically relevant for the protection of national interests. It is no coincidence that this objective, as will be explained in the following pages, is contained in the new National Cybersecurity Strategy in Italy (see Chap. 6), and it is based on a productive coordination among the actors involved in the institutional ecosystem as described in Fig. 4.4.

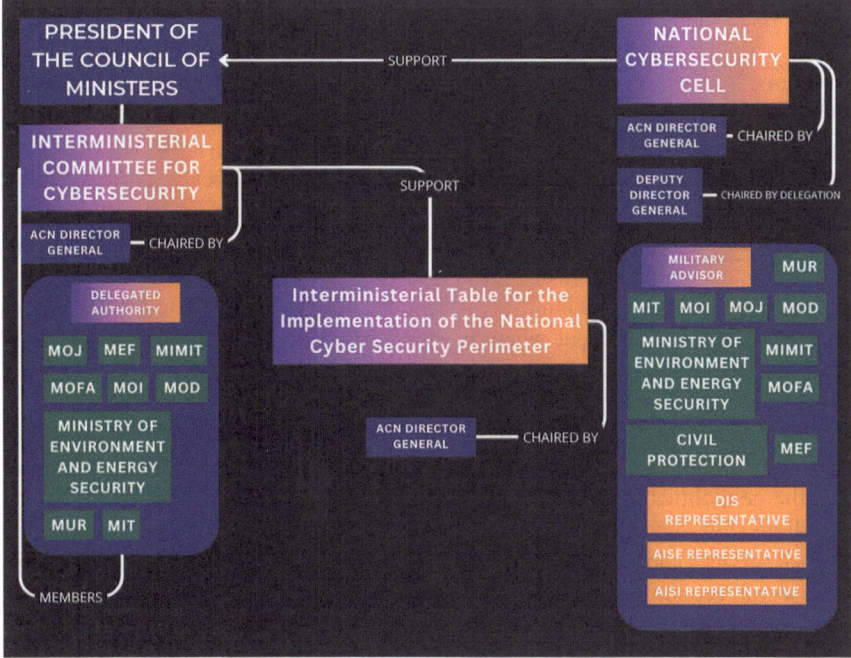

Fig. 4.4 Institutional entities involved in the present Italian cybersecurity governance. (Source: elaborated by author, Luigi Martino (2024))

4.7 Conclusions

More than ten years after the approval of the first Italian cybersecurity governance in 2013, it is possible to draw some empirical observations. The "Monti Decree", on the one hand, has marked a clear distinction between "before and after" the presence of an architecture entirely dedicated to Italy's cybersecurity, the shortcomings of the policies have become evident from the outset: overlapping competent authorities, diffusion of responsibilities, decision-making opacity, heterogeneity of the actors involved, and lack of a dedicated budget for implementing strategic objectives. All these elements, combined with the influences of international initiatives (i.e., NATO and the EU), prompted Italian political decision-makers in 2017 to embark on a path toward reforming the architecture to address the gaps induced by the 2013 choices.

Although the new operational governance introduced by the "Gentiloni DPCM" reform has yielded positive effects, some architectural dysfunctions persist. For instance, the major weaknesses include: a) an inadequate distribution of responsibilities among different governmental bodies, and b) an ineffective decision-making chain between the political-strategic level (headed by the Prime Minister) and the operational level (where Intelligence services, which report directly to the Prime Minister without any decision-making autonomy). For instance, a macroscopic issue stemming from the "Monti Decree" is the absence of a dedicated budget for activities in the cybersecurity realm. Despite the decision in 2015 to allocate "one-time" 150 million euros, this gap remains pronounced in the "Gentiloni DPCM" with relevant operational and functional repercussions. At the same time, the primary role assigned to the DIS, leading to a simultaneous overlap of inspection competencies, collides with the natural function of Intelligence services, which, by their nature, are required to respect secrecy, rather than sharing information widely with the myriad of actors involved in the national cyber ecosystem. This distorted function of intelligence services has significantly affected the real capacity of the cyber "complex machine" to produce the desired effects. These complexities became even more apparent when, in 2019, Italy decided to initiate a regulatory path, somewhat innovative and complex, with the implementation of the national cybersecurity perimeter (PSNC).

However, the turning point came in 2021 with the approval of the new national architecture born in the wake of the changes dictated by the COVID-19 pandemic and the geopolitical tensions. Owing to these needs, Italy has decided to make a paradigm shift, considering cybersecurity no longer the exclusive domain of intelligence services and national security but a fundamental element to ensure the political, social, economic, and industrial resilience of the country. In this context, the National Cybersecurity Agency was also established, which is designated not only as the national Authority in cybersecurity, but also as the single point of contact at the European level and a reference entity at the international level (see Chap. 6).

Table 4.1, reassume the main steps implemented by Italy in the period of time between 2013 and 2021.

Table 4.1 Main policy and governance initiatives summary box between 2013 and 2021

Year	Event	Description
2013	Prime Ministerial Decree of January 24, 2023, "Monti Decree"	The Monti Decree of 2013 marks a crucial moment in Italy's approach to cybersecurity, introducing for the first time a national architecture and dedicated governance to address the challenges of cyberspace
2013	National Strategic Framework for Cyberspace—National Plan for Cyber Protection and Information Security	These documents represent the long-term strategy for 2013–2017 and the short-term operational document, focused on the 2-year period 2014–2015, aimed at identifying the objectives to be achieved and the lines of action to be implemented
2015	Directive of the Prime Minister "Renzi Directive"	On August 1, 2015, the Prime Minister issued a directive aimed at strengthening the national cybersecurity architecture, with the goal of quickly aligning Italy's cyber defenses with those of major international partners and improving the efficiency and effectiveness of the decision-making process
2016	Cybersecurity architecture review	In July 2016, the European Parliament adopted the NIS Directive, which mandated member states to adopt measures to increase the security levels of networks and information systems by May 2018 within the European Union
2016	European Parliament Directive on the security of network and information systems (NIS)	In July 2016, the European Parliament adopted the NIS Directive, which mandated member states to adopt measures to increase the security levels of networks and information systems by May 2018 within the European Union
2016	NATO Warsaw Summit	Following the Warsaw Summit, NATO recognizes cyberspace as a new operational domain in which NATO must defend itself with the same effectiveness as it does in traditional domains: air, land, sea, and space
2017	Prime Ministerial Decree of February 17, 2017, "Gentiloni Decree" and new National Plan	In February 2017, the Gentiloni Decree was published, establishing new organizational structures for Italy's national cybersecurity architecture, recognizing a significant role for intelligence services. Simultaneously, a new version of the National Plan was introduced, updating measures previously adopted in December 2013
2019	Establishment of the national cybersecurity perimeter (PSNC) and approval of the EU Cyber Security Act	The national cybersecurity perimeter was created by Law Decree No. 105/2019 (as converted into law by Law No. 133/2019) and the relevant implementing decrees, including Decrees of the President of Council of Ministers No. 131/2020 and No. 81/2021, Decree of the President of Council of Ministers of June 15, 2021, and Decree of the President of the Republic No. 54/2021. The EU Cybersecurity Act which strengthens the EU Agency for cybersecurity (ENISA) and establishes a cybersecurity certification framework for products and services.

(continued)

Table 4.1 (continued)

Year	Event	Description
2021	Law Decree No. 82/2021, converted with amendments by Law No. 109/2021 "Containing urgent provisions on cybersecurity" and establishment of the National Cybersecurity Agency	Establishment of the third national cybersecurity architecture and governance and establishment of the ACN as National Cybersecurity Authority

Source: Elaborated by author, Luigi Martino (2024)

References and Additional Readings

Cyber Security National Lab—CINI. 2018. The Future of Cybersecurity in Italy: Strategic focus areas. Available in English. https://www.consorzio-cini.it/images/Libro-Bianco-2018-en.pdf

(DPCM Gentiloni) Decreto del Presidente del Consiglio dei Ministri. 2017, February 17. Direttiva recante indirizzi per la protezione cibernetica e la sicurezza informatica nazionali. Available only in Italian. https://www.sicurezzanazionale.gov.it/sisr.nsf/wp-content/uploads/2017/04/DPCM-17-02-2017.pdf

(DPCM Monti) Decreto del Presidente del Consiglio dei Ministri. 2013, January 24. Direttiva recante indirizzi per la protezione cibernetica e la sicurezza informatica nazionale. Available only in Italian. http://www.sicurezzacibernetica.it/db/[2013]%20Decreto%20PCM%2024%20gennaio%202013%20-%20Direttiva%20recante%20indirizzi%20per%20la%20protezione%20cibernetica%20e%20la%20sicurezza%20informatica%20nazionale.pdf

Gazzetta Ufficiale della Repubblica Italiana. Decreto Legislativo 7 marzo 2005, n. 82. https://www.gazzettaufficiale.it/atto/serie_generale/caricaDettaglioAtto/originario?atto.dataPubblicazioneGazzetta=2005-05-16&atto.codiceRedazionale=005G0104&elenco30giorni=false

———. Decreto Legislativo 13 dicembre 2017, n. 217. https://www.gazzettaufficiale.it/eli/id/2018/1/12/18G00003/sg

———. Decreto Legislativo 18 maggio 2018, n. 65. https://www.gazzettaufficiale.it/eli/id/2018/06/09/18G00092/sg

———. Decreto Legge 21 settembre 2019, n.105. https://www.gazzettaufficiale.it/atto/serie_generale/caricaDettaglioAtto/originario;jsessionid=g6v+BHJq0BVT7oc+fikU6A__.ntc-as1-guri2a?atto.dataPubblicazioneGazzetta=2019-09-21&atto.codiceRedazionale=19G00111&elenco30giorni=true

Italian Presidency of the Council of Ministers. The National Plan for Cyberspace Protection and ICT Security, 2013. Available in English. https://www.sicurezzanazionale.gov.it/sisr.nsf/wp-content/uploads/2014/02/italian-national-cyber-security-plan.pdf

———. The Italian Cybersecurity Action Plan, 2017. Available in English. https://www.sicurezzanazionale.gov.it/sisr.nsf/wp-content/uploads/2019/05/Italian-cybersecurity-action-plan-2017.pdf

———. National Strategic Framework for Cyberspace Security, 2013. https://www.sicurezzanazionale.gov.it/sisr.nsf/wp-content/uploads/2014/02/italian-national-strategic-framework-for-cyberspace-security.pdf

Law. 82/2021. Decree-Law No. 82 of 14 June 2021 only in Italian DECRETO-LEGGE 14 giugno 2021, n. 82 Disposizioni urgenti in materia di cybersicurezza, definizione dell'architettura nazionale di cybersicurezza e istituzione dell'Agenzia per la cybersicurezza nazionale URL.

Nato. 2016. NATO Summit Warsaw 2016. URL: https://www.nato.int/cps/en/natohq/events_132023.htm.

NIS. 2016a. Directive (EU) 2016/1148 concerning measures for a high common level of security of network and information systems across the Union.

(NIS DIRECTIVE) Directive (EU) 2016/1148. of the European Parliament and of the Council, 6 July 2016. https://eur-lex.europa.eu/eli/dir/2016/1148/oj

Renzi. 2015. Direttiva del 1 agosto 2015 del Presidente del Consiglio dei Ministri, Only in Italian. URL: https://www.sicurezzanazionale.gov.it/sisr.nsf/documentazione/normativa-di-riferimento/direttiva-1-agosto-2015.html.

De Zan Tommaso. 2016a. Criticità nell'architettura istituzionale a protezione dello spazio cibernetico nazionale, n.117, Approfondimenti, Osservatorio di politica internazionale. http://www.parlamento.it/application/xmanager/projects/parlamento/file/repository/affariinternazionali/osservatorio/approfondimenti/PI0117App.pdf

———. 2016b. Nuova politica di sicurezza cibernetica per l'Italia, Affarinternazionali. http://www.affarinternazionali.it/2016/04/nuova-politica-di-sicurezzacibernetica-per-litalia/

Chapter 5
The National Cybersecurity Perimeter: Regulatory Scope and Limitations

5.1 Introduction

The organizational and regulatory landscape of cybersecurity in Italy was significantly shaped in 2019 with the approval of the Decree-Law of September 21, 2019, No. 105, converted into Law on November 18, 2019, No. 133. The main scope of the Law it was to establish the national cybersecurity perimeter (PSNC). This provision has its roots at the European level in EU Directive 2016/1148 (the aforementioned NIS Directive), whose main purpose was to ensure "a high level of network and information system security within the Union, contributing to raising the common level of security in the European Union" (NIS Directive 2016, 2018). In particular, the need to equip Italy with adequate operability in the context of protecting essential services, stems from the limitations revealed by the transposition and implementation of the NIS Directive into the Italian legal system. In this process, Italy simply decided to adopt the original text of the NIS Directive rigorously and in detail, with a result of transposition of the NIS Directive that we can define as "repetition". Particularly, Italy did not entail either the extension of the subjects included in the definition of operators of essential services (OES), as occurred in other member states of the European Union (e.g., Spain), or the creation of a dynamic mechanism useful for adapting the legislation to evolving cyber threats. The result of this "copy and paste" approach has led to the exclusion of some sectors and strategic operators that fall within the context of essential services. The rationale behind the PSNC legislation is essentially based on the awareness that compromising essential entities would pose a significant threat to national security and citizens' safety. As we will see in the next sections, the PSNC can be defined as an area that encompasses any subject, whether public or private, which provides a "service deemed crucial for the existence of civil, social, and economic activities fundamental to the interests of the State" (PSNC 2019). Therefore, unlike Legislative Decree No. 65 of 2018 of the NIS Directive, the PSNC involves a wide range of subjects, including

L. Martino, *Cybersecurity in Italy*, SpringerBriefs in Cybersecurity,
https://doi.org/10.1007/978-3-031-64396-5_5

those in the governmental sector, such as defense, space and aerospace, energy, telecommunications, economy and finance, transportation, digital services, critical technologies, pension funds and labor organizations. These subjects are required to: (1) adopt specific measures to ensure high levels of security, and (2) compulsorily report to the competent authorities any cyber incidents involving them. In particular, within the context of the PSNC, a further novelty has been the implementation of the assessment and certification process of goods, systems, and computer services intended to be used by the "perimeter actors" through the technical-operational specifications of the National Assessment and Certification Center (CVCN).

To understand the scope of the legislation regarding the PSNC, it is enough to consider that such regulation is also linked to the discipline of Golden Power,[1] which grants explicit powers to the President of the Council of Ministers to oppose acquisitions deemed strategic for national interest in the sectors of defense, energy, transportation, and telecommunications. In summary, the PSNC can be evaluated as a significant step toward protecting national interests, in line with provisions arising from the European Union in the field of cybersecurity. However, as we will discuss in the concluding section, there are significant limitations stemming from the PSNC that, in the medium to long term, risk undermining the operational capacity of the ecosystem affected by the legislation to achieve the expected outcomes. In other words, although the Law is conceived for a "noble purpose" and an ambitious goal, this ambition risks clashing with a complex system, such as that of cybersecurity, where threats evolve rapidly, rendering obsolete, and even harmful, legislation that does not take into account such dynamism. The in-depth analysis of the PSNC is not just an exercise in legislative analysis, but represents a significant contribution to the cybersecurity debate in Italy. Indeed, the national cybersecurity perimeter (PSNC), as Law, offers essential insights for grasping and advancing cybersecurity, given its multifaceted nature, highlighting the need for a holistic approach that not only addresses technological threats but also legal, political, economic, organizational, and social complexities. In other words, the analysis of the PSNC becomes a study model to underline the effectiveness, or deficiencies, of public policies implemented in cybersecurity. Through this examination, on the one hand, we can better understand how legislations can evolve to address both immediate- and long-term challenges in the digital security landscape; on the other hand, we can highlight the limits of political action against the reality of regulatory implementation, especially if adopted with a top-down approach. Ultimately, the reflections and evidence drawn from the analysis of the PSNC Law can also be useful for future legislative actions.

---

[1] The Golden Power refers to a unique authority vested in the Italian government, enabling it to regulate or prohibit (i) foreign direct investments ("FDI") and (ii) corporate transactions involving assets deemed strategically important to Italy. All transactions falling under the purview of the Golden Power mandate must undergo prior notification to the Italian Presidency of the Council of Ministers ("Presidency"). Initially introduced in 2012, the Golden Power was confined to areas such as defense, national security, and critical infrastructure (including transportation, energy, and communications). However, its scope has recently been broadened to encompass additional strategic sectors, such as high technology, fintech, and insurtech, following the enactment of Law Decree 23/2020, which was prompted by the challenges posed by the COVID-19 pandemic.

The lessons learned can guide the development of more flexible, inclusive, and dynamic regulations, which are essential to maintain the ICT integrity and social security in an increasingly interconnected and technology-dependent world. Similarly, through careful analysis, a process of regulatory reform can be initiated, aiming to eliminate the limitations, obstacles, and distortions that emerged during the implementation process.

5.2   Regulatory Content and Operational Provisions of the PSNC

The regulatory and operational framework established by the PSNC, further detailed through subsequent implementing governmental decrees, imposes various obligations on the actors recognized as integral parts of the perimeter. These obligations include, inter alia, the preparation and annual updating of lists of "strategic" ICT assets, which include networks, information systems, and computer services. These lists must be communicated to the competent authorities (i.e., ACN) to allow for verification and inspections. Additionally, there is an obligation to report relevant security incidents to the CSIRT, implement appropriate security measures, and communicate the intention to acquire ICT assets, systems, and services intended for use on strategic assets to specifically designated entities. Furthermore, the PSNC expands the regulatory framework of the "Golden Power" in the context of cybersecurity, aiming to safeguard national companies operating in strategic sectors, particularly in the field of broadband telecommunications networks and technology (i.e. 5G).

In order to operationalize this regulatory framework, a series of implementing Decrees have been approved, as listed below:

1. Decree of the President of the Council of Ministers (DPCM) No. 131/2020 defines the criteria for identifying public and private entities that perform essential functions for the state and fall within the perimeter. The selection of such operators takes place in the areas of space and aerospace, energy, telecommunications, transportation, internal affairs, defense, economy and finance, digital services, critical technologies, and health and welfare entities, through the competent sector authorities.
2. DPCM No. 81/2021 establishes the procedures and modalities for notifying incidents to the CSIRT by the subjects included in the perimeter. This decree imposes stringent notification deadlines, depending on the severity of the incident, and specifies the security measures to be implemented.
3. Presidential Decree (DPR) No. 54/2021 and DPCM of June 15, 2021, outline the methods and procedures related to the operation of the National Assessment and Certification Center (CVCN) and Evaluation Centers (CeVa), as well as specifying the categories of ICT assets, systems, and services subject to communication to the CVCN or CeVa.

4. DPCM No. 92/2022, known as the "Accreditation DPCM," establishes the procedures, requirements, and deadlines for the accreditation of Accredited Test Laboratories, which support the CVCN in verifying the technological security of the entities included in the Perimeter.
5. Law Decree No. 82/2021 (converted with amendments by Law No. 109/2021) redefines the national institutional architecture in the field of cybersecurity, establishing that the National Cybersecurity Agency (ACN) centralizes the PSN functions previously attributed to various government entities, such as the Presidency of the Council, the DIS, the Ministry od Economic Development and the Agency for Digital Italy.
6. Finally, the Determination of the Director-General of the ACN of January 13, 2023, further extends the obligations to report incidents, including a broader taxonomy of incidents and reducing the notification time to 72 h for incidents involving networks, systems, and information services outside the perimeter itself. The following Table 5.1 summarizes the major regulatory and operational initiatives related to the PSNC.

Table 5.1 Major regulatory and operational initiatives related to the PSNC

The title of the provision	Explanation of content
Decree-Law 105/2019 converted into Law 133/2019 Urgent provisions regarding the perimeter of national cybersecurity	Established in order to ensure a "high level of security of networks, information systems, and computer services of public administrations, public and private entities that are of strategic importance in the national landscape"
Expansion of government powers	The PSNC expands the government's powers, known as "Golden Power," to protect national companies operating in strategic sectors during corporate transactions, with a focus on telecommunications networks
Implementing decrees (DPCM No. 131/2020; DPCM No. 81/2021; DPR No. 54/2021 and DPCM of June 15, 2021; DPCM No. 92/2022)	The implementing decrees related to the perimeter contain criteria for identifying subjects within the perimeter, incident notification procedures and security measures, operation of the CVCN and CeVas, and accreditation of Accredited Test Laboratories
DL No. 82/2021, converted with modifications by Law No. 109/2021	This decree redefines the national institutional architecture for cybersecurity, creating the National Cybersecurity Agency (ACN). The ACN takes central responsibilities for the security of networks and information systems
Determination of the Director-General of the ACN	The Determination of the Director-General of the ACN dated January 13, 2023, which extends the incident notification obligations to include a broader range of situations and reduces the notification time to 72 h for incidents outside the perimeter

Source: Elaborated by the author, Luigi Martino (2024)

5.3 The PSNC Approach: Legislative Benefits and the Risks of "Naming and Shaming"

The approach adopted by the Italian government through the establishment of the PSNC represents a remarkable example of how a state actor shapes its posture to address the complex reality of cybersecurity. The law, based on the need to protect essential services, revolves around the establishment of a perimeter within which all actors, regardless of their legal status (public or private), are included if they are holders, managers or supplier of these services. In fact, the law on the PSNC, as indicated in Article 1, is based on the vital need to protect networks, information systems, and computer services essential for the functioning of the state and for the maintenance of civil, social, and economic activities. This objective is situated in a context where "digital security is no longer an option but an imperative necessity for the survival and efficiency of the State itself" (Interview, Rome 2024).

The obligation to exchange information and notify incidents or cyberattacks - a central element of the law - imposes transparency and constant communication between the entities, both public and private, involved in the perimeter and the national Authorities. This approach is justified not just to enable swift and effective responses to threats to national security but also to foster ongoing enhancement of security practices. It aims to create synergy between the entities within the perimeter and the public apparatus responsible for supervision and enforcement. The obligations imposed on the entities included in the perimeter, such as the preparation and updating of lists of ICT assets and the notification of any incidents, aim to create a proactive and osmotic security environment between perimeter entities and supervisory and operational intervention entities.

A fundamental aspect of the law is the obligation to provide information, which requires perimeter entities to document the current situation and communicate the lists of ICT assets and services to the supervisory authority. In this perspective, a service can be categorized as essential if:

- It performs instrumental support activities for the vital functions of the state apparatus.
- It contributes to the exercise and protection of fundamental rights.
- It is indispensable to ensure the continuity of supplies, as well as to optimize infrastructure and logistics.
- It involves research activities and is part of productive realities, both in high technology and in other sectors of economic and social relevance, with particular attention to national strategic security, competitiveness, and the development of the country's economic fabric.

As stated before (see Chap. 4), the implementation of the PSNC also naturally justified the creation of the National Cybersecurity Agency in 2021, which monitors and supervises compliance obligations. From a governance perspective, the PSNC is based on collaboration among various actors: public and private entities, providing essential services to the state, the Interministerial Committee for the Security of

the Republic (CISR), the competent authority (i.e. the ACN), and the National
Evaluation and Certification Center (CVCN) which is also included in the competencies of the ACN. To ensure an inter-institutional interaction among the administrations involved in implementing the national cybersecurity perimeter, the
Interministerial Table for the implementation of the PSNC was established (Article
6 of the Decree of the President of the Council of Ministers No. 131 of July 30,
2020) at the National Cybersecurity Agency, chaired by its Director-General. This
body provides support to the Interministerial Committee for Cybersecurity (CIC),
particularly regarding the identification of essential functions and services of the
state and entities to be included in the perimeter. From an operational perspective,
the National Evaluation and Certification Center (CVCN) plays a critical role in the
PSNC structure, as it conducts technological screening, ensuring that hardware and
software used by perimeter entities are free from vulnerabilities and compliant with
security standards. The role of these bodies is crucial to ensure that security measures meet the needs of the national digital landscape. In fact, the scope of the PSNC
is both geographical and sectoral. Geographically, it extends to the entire Italian
national territory, involving critical infrastructures and essential services. Sectorally,
it encompasses a wide range of areas considered vital for national security, such as
telecommunications, energy, transportation, and digital services. From a regulatory
standpoint, as summarized in the table below, the law imposes specific obligations
on entities that are part of the perimeter, including the annual updating of lists of
ICT assets and the timely notification of significant cyber incidents and attacks.
While these requirements are vital for security, they introduce considerable operational and bureaucratic complexity. This obligation of effective execution of tests on
hardware and software (as required by DPR No. 54) according to experts "could
slow down digitization processes, highlighting a potential conflict between security
and operational agility" (Interview, Rome 2024).

As described in Table 5.2, the obligations imposed on entities included in the
PSNC should translate into concrete and measurable actions, which also entail costs
in terms of personnel and technology that need to be allocated to ensure compliance
with regulations. For instance, the preparation and updating of lists of ICT assets
and the notification of any incidents constitute an operational framework where
transparency and accountability "become vital tools for effective digital security"
(Interviews, Rome 2023, Milan 2024).

To understand the impacts caused by the PSNC Law on the reference ecosystem,
the Annual Report presented to Parliament by the ACN is an extremely useful primary resource (ACN Annual Report 2024). Indeed, according to the 2024
Report (related to the events of 2023) in 2023, the Agency conducted several meet-
ings "with entities subject to the Perimeter norms, which saw changes throughout
the year, concluding with a total of 118 entities" (ACN Annual Report 2024).
Furthermore, according to Article 371-bis of the Criminal Procedure Code
(Legislative Decree No. 36/2023) the Agency is required to transmit relevant data,
information, and updates to the National Anti-Mafia and Anti-Terrorism Prosecutor
to support the coordination and impetus of investigations into serious cybercrimes.
Regarding the impact on Public Administration, the new Public Contracts Code,

Table 5.2 Obligations of the PSNC and description

Obligations of perimeter entities	Description
Incident notification	Prompt reporting of incidents that may affect "ICT assets within the perimeter" is required. This process ensures immediate sharing of information with institutions responsible for the prevention, preparation, and management of cyber threats, such as the NSC and the CSIRT, both under the jurisdiction of the DIS
Adoption of security measures	It is mandatory to implement security measures related to the organization, processes, and procedures for "ICT assets within the perimeter," including aspects related to ICT procurement
Technological screening of ICT procurement	Before acquiring ICT technologies belonging to specific categories for "ICT assets within the perimeter," it is necessary to inform the National Evaluation and Certification Center (CVCN), located at the Ministry of Economic Development. Within a maximum period of 60 days, the CVCN can conduct preliminary checks and impose conditions and hardware and software tests (the latter must be completed within an additional timeframe of 60 days). This procedure is managed in collaboration with the Evaluation Centers (CV) of the Ministries of the Interior and Defense, which deal with the supply of goods, systems, and ICT services intended for their respective networks, systems, and IT services

Source: Elaborated by author, Luigi Martino, 2024 based on (PSNC 2019)

specifies, among the provisions related to procurement criteria (Art. 108), that "contracting authorities, including central purchasing bodies, must always consider cybersecurity elements in the procurement of IT goods and services, especially when the use context involves protecting strategic national interests. In such cases, the contracting authority must set a maximum cap for the economic score (within a 10% limit), emphasizing the importance of assigning appropriate weight to cybersecurity technical-quality profiles over economic ones" (ACN Annual Report 2024). Moreover, at institutional level, the Perimeter Table met three times (in February, October, and December 2023) to deliberate on proposals to update the list of Perimeter entities (ACN Annual Report 2024). Regarding the effects of the Determination of the General Director (see table 5.1) the "Taxonomy of incidents that must be notified" and related obligations - initially only required for incidents impacting assets used in the PSNC (so-called ICT goods) - were extended to incidents affecting networks, information systems, and IT services outside the Perimeter [...] but still pertaining to entities included within it [...] but the introduced provision stipulated a notification deadline within 72 hours, therefore less stringent than those set for Perimeter assets (fixed at 1 hour and 6 hours, depending on the incident's severity)" (ACN Annual Report 2024). Regarding notifications analyzed in 2023 "of the 577 received, 12 concerned annual plans and related updates on 5G technology (pursuant to Art. 1-bis) and 565 concerned the matters referred to in the aforementioned Articles 1 and 2. Additionally, 150 pre-notifications were received. In this context, the Agency was involved in assessing all "5G notifications," and in about 30% of "traditional notifications" that presented cybersecurity profiles. It also examined 23% of the pre-notifications" (ACN Annual Report 2024). All of these data show the intrinsic complexity of the matter and the regulatory landscape emerging from the PSNC impacting the national ecosystem. In this context, the incident notification process, to and from the national CSIRT, plays a fundamental role. Specifically, this process imposes a high level of informational awareness, vigilance, and responsiveness on the involved entities to share malicious events aligned with the obligations. However, this objective must be interpreted in light of the obligations and practical (and bureaucratic) challenges that the Law entails. In particular, while the PSNC Law represents a significant step toward a more systematic and structured approach to cybersecurity in Italy, one of its effectiveness limits is the need to balance cybersecurity with operational flexibility, and the speed of evolving cyber threats. Indeed, although the law aims to implement a system based on the balance between operational efficiency and security rigor, the most evident risk is posed by the moral hazard, due to the need to protect one's personal and organizational reputation, from a potential "naming and shaming" risk arising from admitting a cyberattack.

5.4   Evaluating the Challenges and Opportunities of the PSNC Framework

Despite the numerous benefits offered by the implementation of the PSNC, this regulatory initiative entails some challenges, particularly in terms of operational complexity. While the requirements ensure constant monitoring of critical infrastructure, they also introduce a level of complexity and responsibility that requires careful management by the involved entities. The law emphasizes the importance of communication and transparency between private and public entities and state authorities. This aspect is not only crucial for a timely response in case of incidents but is also a key factor in the continuous improvement process of security measures. Additionally, the law establishes a framework of accountability, where the lack of timely and accurate communication can lead to significant sanctions. This structure demonstrates a holistic approach to cybersecurity, which goes beyond mere protection of infrastructure and also encompasses the verification and certification of ICT products. Table 5.3 summarizes the scope of the PSNC by defining who it applies to, what it applies to, and what the regulations entail.

The PSNC law is based on a delicate balance between the need for robust protection and the reality of a rapidly changing digital environment. The entities involved must navigate through a complex regulatory landscape, fulfilling obligations that can be burdensome but essential for overall security. Another critical aspect is the obligation to communicate for the procurement or acquisition of ICT goods (See ACN Annual Report 2024). This obligation implies a detailed evaluation and approval by the competent Autority, a procedure that ensures the absence of vulnerabilities in the ICT products and services used. However, this process can also pose

Table 5.3 Summary of the scopes of the PSNC based on who and what it applies and what regulations entail and the concept of emergency powers

To whom it applies	National entities, both public and private, that utilize networks, information systems, and computer services to perform essential functions of the state or provide critical services for the maintenance of activities fundamental to society, the economy, or the interests of the state
What it applies to	This concerns the networks, information systems, and computer services of the entities included in the "perimeter." These are activities whose malfunction, interruption (even partial), or misuse could compromise national security
What it entails	The obligation to report incidents, thereby ensuring rapid information sharing with the institutions responsible for the prevention, preparedness, and management of cyber events, particularly the National CSIRT. The adoption of security measures covering organizational, procedural, and process-related aspects, including those pertaining to the procurement of information and communication technology (ICT). A technological review process for ICT procurements in specific categories intended for assets included within the "perimeter." In this process, the entity intending to make such acquisitions must inform the National Evaluation and Certification Center (CVCN) at the National Cybersecurity Agency or the Evaluation Centers (CV) at the Ministries of the Interior and Defense, which, within a maximum period of 60 days, may conduct preliminary assessments and impose conditions and hardware and software tests. The technological screening process is divided into three phases: • Preliminary checks (with a deadline of 45 days extendable by an additional 15 days) • Preparation phase for test execution • Execution of hardware and software tests (activity with a deadline set at 60 days) Inspection and sanction activities carried out by the National Cybersecurity Agency
Emergency powers	In the presence of a serious and immediate risk to national security stemming from the vulnerability of networks, information systems, and computer services, the Prime Minister—following a decision by the Interministerial Committee for the Security of the Republic (CISR)—may, if necessary and only for the time strictly necessary to eliminate or mitigate the specific risk, fully or partially deactivate equipment or products used in the networks, systems, or for the provision of the affected services

Source: Elaborated by author, Luigi Martino, 2024 based on (PSNC 2019)

challenges in terms of costs, timing and operational agility for the entities involved, especially when it comes to acquisitions crucial for daily operations. Additionally, the operational limit posed, for instance by the real availability of access to the source code of the software, must also be considered in this context.

The issue of administrative and criminal sanctions introduced by the regulation, published in June 2021, which establishes "obligations and sanctions to ensure the security of networks, information systems, and computer services" (DPCM 14 aprile 2021, n. 81) as described in the table below, must also be considered (Table 5.4).

In this context, as stated by the PSNC law, sanctions for breaching these obligations include administrative fines ranging from €250,000 to €1,800,000, depending

Table 5.4 Obligations and sanctions derived by the PSNC

Obligation	Description
Incident notification	Entities within the cybersecurity perimeter must report incidents to the Italian CSIRT within specified timeframes, depending on severity: within 6 hours or within 1 hour
Operators of essential services	Operators of essential services in critical sectors such as energy, transportation, banking, and healthcare must comply with the same notification obligations
Technical report	In response to requests from the Italian CSIRT, the entities involved must provide a technical report within 30 days, detailing the significant aspects of the incident and the actions taken to resolve it
Selection of qualified suppliers	Public administrations must select qualified suppliers through AGID for IT services

Source: Elaborated by the author, Luigi Martino 2024, based on (PSNC 2019)

on the specific violation (PSNC 2019, art. 1, c. 9, letters a and b). Looking ahead, although the PSNC law stands as a fundamental pillar in Italy's national security strategy, its effectiveness and agility will hinge on its ability to adapt to new challenges and threats arising from cyberspace. A detailed analysis of this legislation reveals distinct strengths. The law stands out for its ability to promote a cybersecurity culture that extends beyond sectoral boundaries, engaging a variety of actors in a unified defense front. This approach not only enhances the protection of critical infrastructure but also raises overall awareness of digital security. Furthermore, through this initiative, Italy positions itself as a model in the international cybersecurity landscape, promoting high standards and innovative practices. However, the effective and efficient implementation of the PSNC is fraught with challenges. It is clear that cybersecurity requires a dynamic and flexible approach capable of evolving alongside technologies and attackers' tactics. However, the long-term success of public policies in the cyber context depends on the ability to balance security, agility, innovation, and engagement of stakeholders affected by the law. The obligations imposed by the PSNC law represent a crucial step toward creating a more proactive and resilient digital security environment. These obligations not only require entities within the perimeter to keep their ICT asset lists updated but also mandate timely notification of any cyber incidents. This aspect is crucial to ensure that any vulnerabilities or attacks are swiftly identified and managed, reducing the risk of significant damage to critical infrastructure and national security. However, the practical realization of these obligations poses challenges. The rigor and specificity required in documentation and communication can be burdensome, especially for small and medium enterprises that may lack the resources or expertise to meet these complex requirements. This aspect, combined with significant legal responsibilities and a framework of sanctions for entities failing to meet these obligations, adds up to the complexity and costs of implementation. Additionally, the law must address the need to balance responsibilities among the various entities involved. In particular, the burden on smaller organizations and the lack of incentives for updating security practices pose significant obstacles. Furthermore, the law must grapple with the risk of becoming obsolete in a rapidly evolving digital environment, a reality that requires constant commitment to maintaining the effectiveness of the security perimeter. The challenge that emerges is twofold: on the one hand, ensuring the integrity and resilience of systems and, on the other hand, guaranteeing that the security perimeter is implemented effectively and in line with the dynamic needs of society and the national economy. The complexity and the need for effective and efficient practical integration can thus emerge as the Achilles' heel of the PSNC. The need to harmonize different regulations and practices across sectors can cause delays, making the goal of responding quickly to threats a challenging one to achieve. The risk of obsolescence in the face of increasingly sophisticated attack tactics and rapid technological evolution is a gap that threatens to undermine the PSNC's real capacity. In addition to this, despite its innovative potential, the need to harmonize different regulations and practices can lead to delays and difficulties in the rapid response to threats. Furthermore, the regulatory framework clashes with the reality of compliance costs and burdens, particularly burdensome for smaller entities (i.e. SMEs and start-up). Regarding the economic impacts, it is clear that for the case of private entities,the costs can fall on consumers, and in the case of public administration on taxpayers. The lack of adequate differentiation between public and private entities of different sizes and the absence of tax and financial incentives for adherence and updating of security practices thus risk reducing the overall effectiveness of the PSNC.

5.5 Conclusions

The enactment of the PSNC represents a strategic turning point in Italian legislation regarding cybersecurity; its introduction signals a mature national awareness of the urgency to address the challenges posed by cyberspace. This legislation not only aims to protect critical infrastructure but also to establish an integrated security paradigm that includes both public and private entities. It is a proactive response to the growing threats in cyberspace, recognizing the need for a holistic and cross-sectoral approach to national security. While outlining the "perimeter," policymakers have faced significant complexity—such as the need to apply an inter-sectoral coordination approach, harmonizing diverse regulations and operational practices—this challenge of complexity will increasingly be crucial to measuring the actual effectiveness of the PSNC against expected outcomes. In particular, the law carries a fundamental responsibility: the need to ensure that the regulatory and punitive framework does not have the opposite effect of overwhelming perimeter operators with stringent, economically inefficient demands that are not keeping pace with rapidly evolving and unpredictable threats. This aspect requires continuous synergy between the supervisory authority and the regulated operators, aspects that impose a significant burden in terms of resources and strategic commitment. The Law's primary goal is to safeguard critical infrastructure and protect the operational continuity of the country's vital services from cyber threats, through an approach of shared security and responsibility capable of integrating both the public and private sectors. Last but not least, the PSNC law has also introduced significant changes in national governance, aiming to ensure an appropriate architecture for national cybersecurity, which has led to the creation of the National Cybersecurity Agency.

References and Additional Readings

ACN Annual Report. 2024. 2023 Year in Review. URL: https://www.acn.gov.it/portale/documents/20119/446882/ACN_Review_2023.pdf.

(Cybersecurity Act) Regulation (EU) 2019/881 of the European Parliament and of the Council of April 17, 2019: On ENISA, the European Union Agency for Cybersecurity, and on Cybersecurity Certification for Information and Communication Technology (ICT) Products and Services, and Repealing Regulation (EU) 526/2013.

Decree-Law October 18, 2012, No. 179: Additional Urgent Measures for the Country's Growth.

(PSNC 2019) Law September 21, 2019, No. 105: Urgent Provisions on the National Cybersecurity Perimeter and the Regulation of Special Powers in Strategic Sectors.

Directive (EU) 2022/2555 of the European Parliament and of the Council of December 14, 2022: On Measures for a High Common Level of Cybersecurity in the Union, Amending Regulation (EU) No 910/2014 and Directive (EU) 2018/1972 and Repealing Directive (EU) 2016/1148 (NIS 2 Directive).

(NIS Directive) Legislative Decree May 18, 2018, No. 65: Implementation of Directive (EU) 2016/1148 on Measures for a High Common Level of Security of Networks and Information Systems in the Union.

Prime Ministerial Decree February 17, 2017: Directive Providing Guidelines for Cyber Protection and National Cybersecurity.

PSNC. 2019. Law September 21, 2019, No. 105: Urgent Provisions on the National Cybersecurity Perimeter and the Regulation of Special Powers in Strategic Sectors.

——— August 8, 2019: Provisions on the Organization and Functioning of the Computer Security Incident Response Team—CSIRT Italia.

——— July 30, 2020, No. 131: Regulation on the National Cybersecurity Perimeter, Pursuant to Article 1, Paragraph 2, of Decree-Law September 21, 2019, No. 105, Converted, with Amendments, by Law November 18, 2019, No. 133.

Regulation (EU) 2021/887 of the European Parliament and of the Council of May 20, 2021: Establishing the European Cybersecurity Competence Centre within the Framework of the Industrial, Technological and Research Areas and the Network of National Coordination Centers.

Chapter 6
Insights into the National Cybersecurity Agency

6.1 Introduction

In the midst of the ever-changing cyber landscape and the gaps in its national coordination structure, Italian policymakers have increasingly realized the need to rethink the approach to cybersecurity governance. Therefore, starting from the first two decades of the twenty-first century, the main objective of Italian policymakers has been to ensure the protection and resilience of critical infrastructures and vital services through a harmonized institutional framework. This need has been realized through a broader renewal of the governance process, as well as a more effective inter-institutional coordination mechanism. In response to these needs, the National Cybersecurity Agency (ACN) was created, established by Decree-Law No. 82 of June 14, 2021, which precisely redefined "the national architecture of cybersecurity with the aim of rationalizing and simplifying the system, enhancing aspects of Italian cybersecurity and resilience" (see Chap. 3).

In this new institutional framework, the ACN has been tasked to play the role of the national Authority for cybersecurity. This assignment includes, inter alia, the responsibility to protect both Italian interests in cyberspace and to ensure effective assistance for maintaining the digital resilience of the entire country. At the same time, this Agency also holds the task of preventing and mitigating cyberattacks and promoting the achievement of Italian technological autonomy, as well as spreading the culture of cybersecurity. Among the other main responsibilities of the ACN is the implementation of the goals stated in the National Cybersecurity Strategy. From a governance perspective, the creation of the ACN has entailed a "Copernican revolution", as it has involved a complete curtailing of the competencies, previously allocated to various institutional actors. The first downsizing affected the role of intelligence services (i.e., DIS), which transferred cyber functions to the ACN starting from 2021. Additionally, the ACN has absorbed the competencies of the Ministry of Economic Development (MiSE), including the responsibility related to the

National Evaluation and Certification Center (CVCN). In summary, starting from 2021, the ACN has become the competent national body in the context of cyber security and resilience.

Throughout this chapter, we will carefully examine the path that led to the creation of the ACN and evaluate "why" and "how" this initiative represents a crucial step in addressing the increasingly complex challenges of cybersecurity in Italy. The fundamental question is: does the ACN represent an effective response to the growing cyber threats? In our attempt to answer this question, we will examine the strengths and weaknesses of the ACN. Additionally, we will seek to determine whether the centralization of cybersecurity governance in Italy, achieved through the establishment of an ad hoc Agency, addresses previous shortcomings, or if the stated objectives actually indicate a *vaste programme* for the new architectural structure.

6.2   Exploring the Mission, Objectives, and Structural Organization of the National Cybersecurity Agency

The establishment of the ACN was a bold and innovative move by Italian policymakers. However, it is important to critically examine the reasons behind this choice and the potential risks associated with it. The fundamental questions may be the following: why was the ACN created? What drove Italy to concentrate so much power and responsibility in a single entity?

Officially, the establishment of the ACN was motivated by the need "to ensure a coordinated and cohesive response to increasingly sophisticated and pervasive cyber threats capable of targeting Italian economic, political, and social interests" (Interviews Rome and Milan, 2024). The history of the ACN begins with the approval of Decree-Law No. 82 on June 14, 2021, subsequently converted into Law No. 109 on August 4, 2021 (see Chap. 4).

According to this reform, the Prime Minister was assigned the high direction and overall responsibility for cybersecurity policies. Additionally, the Interministerial Committee for Cybersecurity (CIC) was established, which has advisory, proposal, and oversight functions regarding cybersecurity policies. The ACN also manages additional structures such as the National Evaluation and Certification Center (CVCN), the Computer Security Incident Response Team (CSIRT) Italy, and the Cybersecurity Nucleus (NCS), which deal with prevention and preparation for crisis situations and activation of alert procedures. Furthermore, an interministerial table for the implementation of the Perimeter of national cybersecurity (PSNC) is envisaged (see Chap. 5).

The ACN is therefore established with the main objective of safeguarding national interests in the field of cybersecurity, with the aim to enhance and maintain the resilience of the entire country" (Law 82/2021). Moreover, the Agency is endowed with regulatory, administrative, patrimonial, organizational, accounting, and financial autonomy, within the limits set out by the Law. The main objective of the Agency is to contribute to the "implementation of common actions, aimed at guaranteeing levels of cybersecurity and cyber resilience which can foster the country's digital development" (Interview, Rome 2024). It also offers specific training courses for the development of the workforce in the cybersecurity sector and supports awareness-raising campaigns (Interview, Rome 2024). At governmental level, the Agency, supports the President of the Council of Ministers in the field of cybersecurity, more specifically in crisis prevention and preparedness and for the activation of alert procedures.

During the period 2021–2023, the ACN focused on various preliminary and functional activities, such as the development of internal regulations and procedures, the creation of committees and bodies, and the selection of necessary personnel. Parallel to its "construction process", the Agency has quickly become an active player in the institutional landscape, with specialized competencies in cybersecurity. However, this position should be interpreted in an inter-institutional context, in collaboration with other functions attributed to various public entities. In particular, cooperation includes interaction with inter-institutional liaison offices, such as the Interministerial Committee for Cybersecurity (CIC) at the political level, and the Cybersecurity Unit (NCS) at the technical and operational level. The CIC plays a significant role in providing guidance, advice, and oversight in cybersecurity matters at the political level, including monitoring the implementation of the National Cybersecurity Strategy and providing opinions on acts necessary for the Agency's operations (see Chap. 4).

The ACN is also designated as the point of contact for networks and information systems security, as well as the National Certification Authority for Cybersecurity and the National Coordination Center for Cybersecurity in the industrial, technological, and research fields (ACN Annual Review 2023). This implies that the ACN is committed to: (1) pursuing strategic technological autonomy, at both national and European levels and (2) is called upon to collaborate closely with the national industry, the academic and research communities. From an operational perspective, the role played by the ACN in the Italian institutional and regulatory context can be divided into four main categories of competencies: (i) active administration and technical operational activities aimed at ensuring security and resilience, (ii) coordination and liaison functions between national and international administrations, (iii) supervisory activities, and (iv) promotion of research and technological development in the field of cybersecurity.

In this new governance framework, as emerges from the following organizational chart (Fig. 6.1), the ACN, in line with the public-public partnership approach (see Chap. 2), operates synergistically with other institutional functions, such as the law enforcement forces (continuing to handle criminal cyber investigations), while the Intelligence Agencies (regarding cyber intelligence operations or cyber threats that can affect national security).

As for its internal structure, starting from 2021, the agency has adopted around 40 implementation measures to organize its functions and activities. It gradually activated services, divisions, and other units necessary for the effective execution of

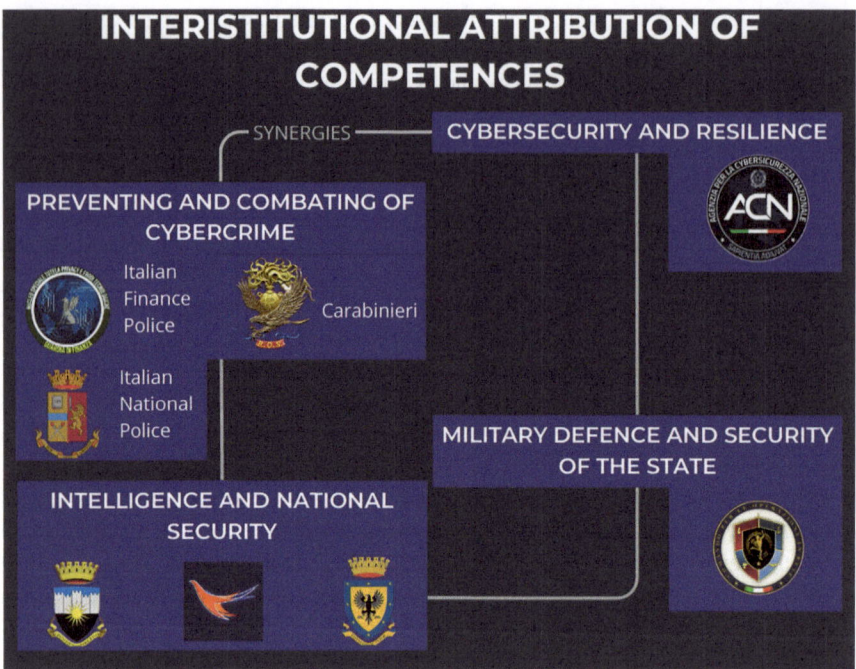

Fig. 6.1 The public-public partnership approach of the Italian National Cybersecurity Agency. (Source: elaborated by the author, Luigi Martino (2024))

priority tasks. The ongoing selection of personnel, during the first biennium 2021–2023, allowed for the redistribution of responsibilities, aligning them with the agency's growth both in terms of organizational structure and personnel size. According to the Annual Report (ACN Annual Report 2023), by the end of 2022, the Agency's internal structure included five services, 16 divisions (nine of which were of greater complexity), two mission units, one unit supporting the Director-General, and one project team. From an organizational perspective, the ACN can be represented as described in Fig. 6.2.

This structure allows the ACN to act as a crucial actor in the cybersecurity landscape in Italy, actively collaborating with various governmental entities to enhance national cybersecurity and resilience. In this collaborative context, as stated before, the Cybersecurity Unit (Nucleus for Cybersecurity, NCS) plays a fundamental role, being responsible for coordinating and supervising cybersecurity activities at both technical and operational levels (see Chap. 4). In other words, from a governance standpoint, the NCS plays a central role in the prevention, preparation, and response to cyber crises in Italy. The role of the National Cybersecurity Unit (NSC) can be summarized as follows:

- *Permanent coordination*: The NCS operates as a permanent body within the ACN, under the guidance of the Agency's Director-General. Its mission is to

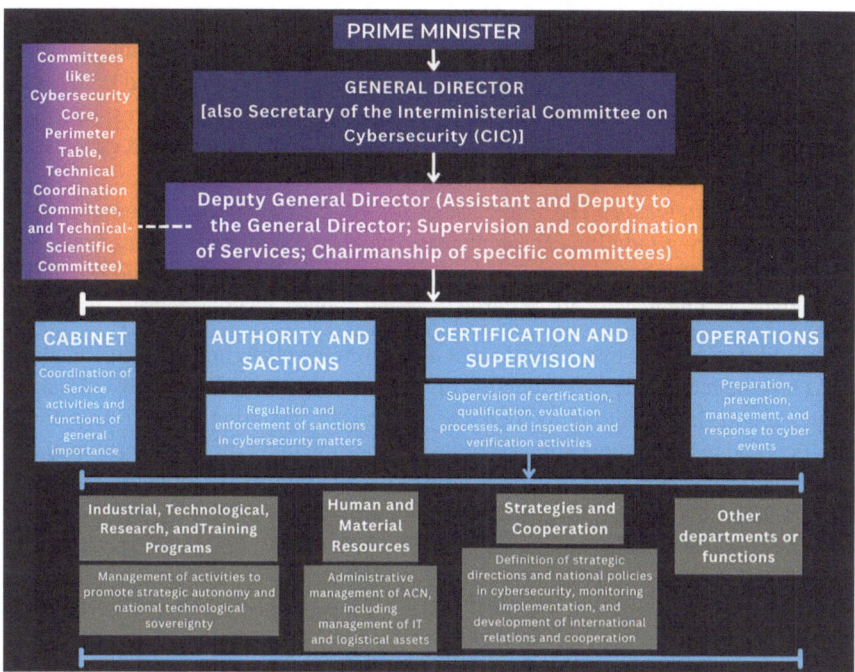

Fig. 6.2 The organizational structure of the Italian National Cybersecurity Agency. (Source: elaborated by the author, Luigi Martino (2024))

support the Prime Minister in all cybersecurity-related matters, with a particular focus on prevention and preparation for cyber crises.

- *Composition*: The NCS comprises key representatives, including the Prime Minister's Military Advisor, representatives from the Department for Information Security (DIS), the Agency for External Information and Security (AISE), the Agency for Internal Information and Security (AISI), and representatives from various ministries involved in the Interministerial Committee for Cybersecurity (CIC). It also integrates classified information management capabilities through a representative from the Central Office for Security (UCSe) within the DIS.
- *Flexible response*: The NCS can be convened in a restricted composition when it comes to specific administrative and security issues, especially those related to cyber crisis management. This flexibility allows it to adapt its composition according to the situation's needs.
- *Exercises and preparation*: In addition to its coordination role, the NCS is actively involved in planning and conducting exercises, often concerning multiple sectors and domains. These exercises serve not only to prepare for cyber crises but also to test and refine established procedures and provide training to the involved personnel.
- *Advisory and policy role*: The NCS has taken on an important role in formulating cybersecurity policies and regulations. It has contributed to the development and

adoption of regulatory measures aimed at strengthening national cybersecurity, such as the diversification of technological products and services to mitigate risks.
- *Rapid response*: The unit's ability to rapidly share information and collaborate with various government agencies and organizations is crucial for responding to rapidly evolving cyber threats. It plays a central role in incident and cyberattack management, ensuring a coordinated and effective response.
- *Regulatory influence*: The influence of the NCS extends to the formulation of cybersecurity regulations and policies, which can have a significant impact on Italy's overall cybersecurity posture. Its role in defining the legal and regulatory framework underscores its importance in the country's cybersecurity strategy.

In addition to these operational functions, given the changing geopolitical conditions and the increasing sophistication of cyber threats, the NCS also provides advisory support in managing potential cyber crises, resulting from international conflicts and technological vulnerabilities. In addition to its specific operational role within the National Cybersecurity Strategy (NCS), as outlined in the ACN Annual Report 2024, the Agency undertook various initiatives in 2023 based on the prerogatives granted by the regulatory framework. Notably, ACN also serves as the National Coordination Centre (NCC), working closely with the European Cybersecurity Competence Centre (ECCC) to support its initiatives aimed at strengthening industrial, technological, and research developments in cybersecurity. In this capacity, ACN helps the ECCC manage funding opportunities related to cybersecurity in industrial, investment, and innovation sectors, as provided by the Digital Europe Programme (DEP) and Horizon Europe (ACN Annual Report 2024).

6.3   Financial Allocation and Strategic Direction in National Cybersecurity Efforts

One of the most important novelties introduced by Law No. 82/2021, compared to past approaches, was also the permanent financial allocation provided to the new national cybersecurity governance held by the ACN. In fact, in order to address the challenges and opportunities related to digital transformation, the policymakers have conferred upon the Agency an important institutional position in governance and a high degree of autonomy. This autonomy allows it to adopt special strategic, organizational, and personnel management approaches. In this perspective, the "ACN Strategic Plan" plays a fundamental role, a document summarizing the main challenges the agency is facing, anticipating possible future scenarios. This plan serves as a guide for planning, governance, and coordination, emphasizing strategic objectives, recruitment needs, personnel training and development, and promoting continuous organizational improvement.

From a financial standpoint, Law No. 82/2021 granted the agency accounting and financial autonomy. In 2022, the first provisional budget was prepared, adopted on June 24, 2022, and the final budget for 2021, approved on December 19, 2022.

The balance sheet as of December 31, 2021, showed a surplus of 1,768,035 euros, while in 2022 revenues totaling 104,048,942 euros were recorded. These resources are intended to support the Agency's present expenses and finance investments, including projects of the PNRR directly managed by the agency or entrusted to other public administrations. Additionally, in 2022, the funds allocated under Article 17, paragraph 7, of Decree-Law No. 82/2021, totaling 2.1 million euros, were utilized to procure goods and services essential for initiating institutional operations and covering logistical expenses for the headquarters (ACN Annual Review 2023). The fiscal landscape in 2023 is also evolving. According to the Annual Review 2024, the initial allocation of €623 million has been set to support the initial functioning of the Agency (ACN Annual Review 2024). This investment is structured around three main pillars: (1) Enhancing the cyber capabilities of Public Administration to secure data and services for citizens. (2) Building cyber resilience across the state by fostering synergies and interconnections for monitoring, sharing information, and responding to cyber events. (3) Developing national technology scrutiny and certification capabilities to assess and certify ICT goods, systems, and services, as part of the activation of the CVCN at ACN. The funds available through the Budget Law for 2023 include the Fund for the Implementation of the National Cybersecurity Strategy and the Fund for Cybersecurity Management, allocated to various administrations responsible for specific interventions. In addition to these dedicated funds, the entities involved can also access other financing sources, including administrative funds and those from the National Recovery and Resilience Plan (PNRR). Specifically, the financial statement for 2022 reported an operating profit of €21,706,431, "which is earmarked for creating a reserve to cover future investment expenses aimed at enhancing the Agency's effectiveness" (ACN Annual Report 2024). Furthermore, the revised economic budget for 2023 and the economic budget for 2024 were adopted on August 1 and October 31, 2023, respectively. In this context, the 2022 financial statement, adopted on April 28, 2023, by the Director General's resolution and approved by the Prime Ministerial Decree on July 6, 2023, following the opinion of the Interministerial Committee for Cybersecurity, recorded an operating profit of €21,706,431 (ACN Annual Report 2024). This amount is designated to establish "a net equity reserve for covering future investment expenses to boost the Agency's effectiveness" (ACN Annual Report 2024). The 2023 Budget Law (Law No. 197/2022) established the Fund for the Implementation of the National Cybersecurity Strategy to finance investments aimed at achieving technological autonomy in the digital realm and improving the cybersecurity levels of national information systems. Additionally, the Fund for Cybersecurity Management ensures economic coverage for operational management activities (ACN Annual Report 2024). The total amount of resources allocated for the period 2023–2027 is nearly €497 million (see graph Fig. 6.3) and are almost all dedicated to financing specific projects essential for achieving the Strategy's objectives for the period 2022-2026 (ACN Annual Report 2024).[1]

---

[1] The total amount of resources allocated for the period 2023–2027 is nearly €497 million (see Fig. 6.3), almost all of which is dedicated to financing specific projects essential for achieving the Strategy's objectives for the period 2022–2026 (ACN Annual Report 2024).

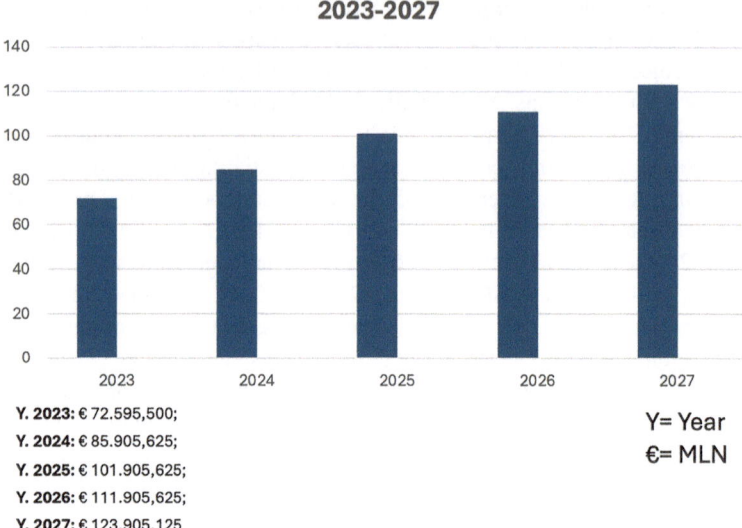

Fig. 6.3 ACN Annual Report 2024

As stated before, the majority of the financial availability is allocated to the ACN to pursue the strategic objectives. To this end, on May 17, 2022, both the National Cybersecurity Strategy 2022–2026 and its Implementation Plan were published containing 82 measures, detailing how to achieve protection, response, and development objectives. Many of these measures promote a "whole-of-society" approach involving the public and private sectors, civil society, academia and research, media, families, and individuals to strengthen Italy's cyber resilience. Therefore, the strategy aims to support a resilient digital transition by implementing tools to assess the security of ICT solutions adopted in the country's infrastructure, including aspects related to national procurement and supply chain. In parallel, the adoption of encryption throughout the life cycle of ICT systems and services is promoted. The strategy also foresees the development of high national capabilities for monitoring, detection, analysis, and management of cyber threats, as well as tools and useful indications to address cyber crises. One of the key objectives of the Strategy is to actively promote a culture of cybersecurity to increase awareness in the public and private sectors, as well as in civil society, regarding cyber risks and threats. Finally, emphasis is placed on strengthening cooperation both at the national level (in terms of partnership between the public and private sectors, between public entities, academia), and at the international level (actively participating in European and international initiatives and promoting bilateral collaborations). Regarding this latter aspect, the Annual Report 2024 highlights that ACN has actively participated in several cybersecurity-related forums at the multilateral level (ACN Annual Report 2024). This includes involvement in multilateral institutions such as the United Nations, NATO, and the OSCE, as well as intergovernmental dialogues like the G20 and G7. A notable example is the "ACN's involvement in the Counter-Ransomware Initiative [...] At this summit, the 50 member states signed a joint declaration committing their public administrations not to pay ransoms to cybercriminals and to provide mutual assistance in the event of ransomware attacks on critical sectors" (ACN Annual Report 2024). Additionally, also in the topic of AI "the Agency endorsed the "Guidelines for Secure AI Systems Development," published by the UK National Cyber Security Centre and the US Cybersecurity and Infrastructure Security Agency" (ACN Annual Report 2024).

Despite these significant developments, one of the main challenges that emerged from the analysis of strategic objectives is the effective involvement of the private sector and civil society in cybersecurity initiatives. Although a "whole-of-society" approach is promoted, there are concerns about the actual ability to foster a bottom-up approach. In fact, one of the elements emphasized by representatives of various private organizations directly affected by the actions implemented by the new Italian cybersecurity governance is "the limited involvement in decision-making processes" (Interviewees Milan and Rome 2023, 2024). Specifically, as we stressed in Chat. 2, a successful cyber ecosystem requires to move beyond a mere "talk of inclusion" approach. Instead, efforts should focus on establishing mechanisms to encourage effective partnerships between the public and private sectors, academic institutions, civil society, and other stakeholders. This requires a constant commitment to information sharing, direct and equal involvement of all the components of the ecosystem, training of highly qualified personnel, collaboration in responding to cyber threats, and fiscal incentives.

According to almost all the interviewees involved in the empirical validation of the observations included in this book,[2] efficiency in the management of financial resources is another critical aspect that requires particular attention. Since significant funding is referenced for the initiatives planned by the ACN, "it is equally essential that funds are used effectively and efficiently. This means that rigorous mechanisms of financial control and accountability must be established to ensure that funds are spent responsibly and in accordance with established objectives" (Interview, Rome 2024). A significant challenge ahead for ACN will be the implementation of the Directive on measures for a high common level of cybersecurity across the Union (NIS2 Directive), approved at the European level in 2023 and adopted by Italy on June 13, 2024. Specifically, the EU Directive 2022/2555 (i.e. NIS2 Directive) builds upon and strengthens the regulatory framework established by the previous NIS Directive (i.e. NIS 1). The NIS 2 incorporates lessons learned from its application in the following areas: (i) Introduction of an identification mechanism for entities deemed important or essential, based on a uniform size-cap rule. This extends the Directive's application to all medium and large enterprises operating in the sectors identified by the Directive. Additionally, central Public Administration entities are included (with discretion left to Member States regarding local administration) and, regardless of size, certain specific categories of enti-

---

[2]The total number of interviewees is 35 individuals who hold various top positions in public or private organizations directly or indirectly involved in the Italian cyber ecosystem. The anonymized interviews were conducted between 2023 and 2024.

ties are also covered. (ii) Expansion of the scope of application, significantly increasing the sectors covered and introducing an "all-hazards" approach to cybersecurity, which includes physical security aspects of ICT infrastructure. (iii) Strengthening of supervisory powers, with more detailed guidelines for defining security measures and increased penalties. (iv) Extension of the functions of national CSIRTs, which will act as trusted intermediaries between reporting entities and ICT product and service providers within the coordinated vulnerability disclosure framework. (v) Crisis management, with the establishment of national frameworks and the institutionalization of EU-CyCLONe for coordinated operational management of large-scale cybersecurity incidents and crises. The transition from NIS to NIS2 requires an update to the regulatory framework across all EU countries, which must implement the new Directive by October 17, 2024. In this context of a "normative deluge", ACN will need to demonstrate its ability to leverage the national ecosystem effectively, while addressing the risk of over regulation.

6.4   Conclusions

The establishment of the ACN in 2021 was primarily driven by political considerations. Italian decision-makers recognized that fragmented responsibilities and ineffective coordination among government agencies could undermine Italy's ability to address cyber threats effectively. Therefore, the creation of the ACN aimed to consolidate the expertise and resources necessary to address cyber challenges synergistically. Secondly, the ACN was established to address regulatory compliance resulting from European demands and to ensure the protection of the interests of all stakeholders, both public and private, in the national cyber ecosystem. The underlying rationale was to ensure mature national cybersecurity, by establishing an Agency dedicated to both operational management and the due diligence on policies and regulations. Furthermore, according to the organizational design of the new governance dedicated to Italian cybersecurity, the ACN is required to play a coordinating and collaborative role between the public and private sectors to comprehensively address cyber challenges.

However, as emphasized by experts and professionals interviewed, concentrating all these responsibilities in a single entity entails both benefits and potential risks. In particular, although the centralization of operational competencies is a significant element in the cyber context, it is also important to ensure that the ACN has the necessary resources and expertise to effectively fulfill its role continuously, beyond the exceptional moment represented by EU' Recovery and Resilience Plan funds. Additionally, centralization could have a detrimental effect within the layered Italian governance context. Therefore, adequate oversight is crucial to avoid a dominant position, or conflicts of interest, that could lead to potential decision-making paralysis. Indeed, although the ACN was born out of the need to address an emerging threat that knows no geographical boundaries, there is a risk that the Agency could become a "bureaucratic monster", burdened by excessive regulatory complexity and slow decision-making. Its operational effectiveness could be called into question if it fails to keep pace with the rapid evolution of cyber threats and the incoming "normative deluge" from European Union.

Another potential weakness lies in the "naming and shaming" approach stemming from regulatory mandates (such as those highlighted in the Chap. 5 regarding the obligations contained in the Perimeter Law). This approach could contradicts the "whole-of-society" concept contained in the national strategy. In this regard, the choice to attribute to the ACN the responsibility for regulatory compliance and the protection of the interests of all stakeholders, both public and private, may seem logical to ensure comprehensive management of cybersecurity issues. However, this concentration of power could lead to a kind of regulatory and control monopoly over cybersecurity, with the potential risk of "anomie" from other actors involved in the national cyber institutional ecosystem, who, without direct involvement, would feel excluded from decision-making processes. Moreover, the private sector may feel trapped and controlled by a government Agency, tasked with regulatory compliance control and auditing, with potential repercussions on business operational continuity.

Finally, the placement of the ACN under the subordination of the President of the Council of Ministers was justified by the importance of cybersecurity as a fundamental part of national security. Nevertheless, this raises a series of political questions that remain open to date. Who ensures that the ACN will not become a tool of political power, used for purposes other than cybersecurity protection? What guarantees are there that the ACN operates independently and remains unaffected by the spoil system? In summary, as some interviewees have emphasized, "it could legitimately be asked whether the concentration of all these responsibilities in a single entity entails risks related to the dimension, organizational structure, and actual capabilities of action of the ACN" (Interviews Rome and Milan, 2023–2024).

One of the most contentious issues regarding the political decision-maker's choice to centralize the roles of coordination, supervision, and operations in a single Agency concerns the evaluation of whether such functions would have been more useful if attributed to independent actors. According to those who advocate for this necessity, this could bring two significant advantages. On the one hand, it could limit the size and functions of the ACN, thus favoring its operational effectiveness. On the other hand, it would allow for a clear distinction between operational activities and those aimed at ensuring regulatory compliance and protecting the interests of all stakeholders. Ultimately, one of the main challenges of the ACN is its "work in progress" phase. As highlighted by the information provided in the ACN's first Annual Report delivered to the Parliament, much of the activities in 2022 were dedicated to building and consolidating the Agency itself. This raises questions about the ACN's full operational capacity and its ability to effectively attract top talent people to the Agency and retain them for the long term, to serve its strategic goals. Collaboration with other institutional actors in interministerial working table or fora (such as the CIC and the NCS) is a crucial element in effectively addressing cyber threats. However, such collaboration can be complex to manage in terms of coordination and communication, between different Agencies, if a public-public partnership approach is not implemented, even before a public-private partnership. Indeed,

although this principle of partnership is written in various strategic and policy documents, a gap in terms of operational collaboration at multiple levels (central and
peripheral) could lead to delays in responses, or competence conflicts, that could
jeopardize the overall effectiveness of cybersecurity governance. From a perspective of democratic checks and balances, regarding the placement of the ACN within
the government organization, it was inevitable to consider it subordinate to the
President of the Council of Ministers, given the importance of cybersecurity as a
fundamental part of national security. However, although the ACN is rightly subject
to the oversight of the Parliamentary Committee for the Security of the Republic
(COPASIR), which implies a high degree of transparency and accountability, this
aspect could also entail greater bureaucratic complexity and the risk of political
interference in the Agency's operations. Furthermore, the ACN's accounting and
financial autonomy require rigorous financial responsibility, and it will be essential
to demonstrate real and tangible results for the entire national ecosystem in exchange
for such autonomy.

References and Additional Readings

ACN Annual Report. 2023. https://www.acn.gov.it/portale/documents/20119/99437/ACN_
    Relazione_2022_ENG.pdf.
———. 2024. https://www.acn.gov.it/portale/en/2023-year-in-review.
Decree of the President of the Council of Ministers June 15, 2021: Identification of Categories
    of ICT Goods, Systems, and Services Intended to be Used within the National Cybersecurity
    Perimeter, in Implementation of Article 1, Paragraph 6, Letter a), of Decree-Law September
    21, 2019, No. 105, Converted, with Amendments, by Law November 18, 2019, No. 133.
———. December 9, 2021, No. 223: Organization and Functioning Regulation of the National
    Cybersecurity Agency
———. December 9, 2021, No. 224: Personnel Regulation of the National Cybersecurity Agency.
———. December 9, 2021, No. 222: Accounting Regulation of the National Cybersecurity Agency.
Decree-Law June 14, 2021, No. 82: https://www.gazzettaufficiale.it/eli/id/2021/06/14/
    21G00098/sg
Directive (EU) 2022/2555 of the European Parliament and of the Council of December 14, 2022:
    On Measures for a High Common Level of Cybersecurity Across the Union, Amending
    Regulation (EU) No 910/2014 and Directive (EU) 2018/1972, and Repealing Directive (EU)
    2016/1148.
Implementation Plan National Cybersecurity Strategy 2022–2026. https://www.acn.gov.it/portale/
    documents/20119/87708/ACN_EN_Implementazione.pdf/4281ece4-908c-cc1b-811e-502841
    fe2351?t=1704814433371
Martino and Soi. 2021. Recovery cyber. L'agenzia vista da Soi e Martino, Formiche, 29.05.2021.
    URL: https://formiche.net/2021/05/agenzia-cyber-pnrr-soi-martino/#content.
National Cybersecurity Agency (ACN) National Cybersecurity Strategy 2022–2026. https://www.
    acn.gov.it/portale/en/strategia-nazionale-di-cybersicurezza
NIS 1 Directive (EU) 2016b/1148 of the European Parliament and of the Council of 6 July 2016
    concerning measures for a high common level of security of network and information sys-
    tems across the Union. URL: https://eur-lex.europa.eu/legal-content/EN/TXT/?uri=celex%
    3A32016L1148
———. 2 Directive on measures for a high common level of cybersecurity across the Union URL:
    https://eurlex.europa.eu/eli/dir/2022/2555.

Chapter 7
Conclusions

The cybersecurity regulatory landscape in Italy is continually evolving. A recent development, enacted just before this book went to press, is Law No. 90 of June 28, 2024, titled "Provisions for Strengthening National Cybersecurity and Cybercrime." This law aims to enhance national cybersecurity and strengthen the resilience of public institutions in Italy. It requires specific public administrations to report cybersecurity incidents within defined timeframes and imposes penalties for non-compliance with corrective measures. However, as in the classic game of cat and mouse, attackers are continually pursued by defenders. To grasp the scale of cyber threats, consider that in 2023, Italy was the third most targeted country in the European Union by ransomware attacks and the sixth globally (ACN Annual Report 2024). That year, the National Cybersecurity Agency (ACN) managed 422 cyber incidents affecting national public institutions, a substantial increase from 160 incidents in 2022. Of these, 85 were classified as significant incidents, compared to 57 the previous year, causing severe disruptions such as system malfunctions and service interruptions. In 2023, the Italian CSIRT identified 3,302 cyber targets, a significant rise from 1,150 in 2022 (ACN Annual Report 2024). The CSIRT Italy handled 1,411 cyber events in 2023, averaging about 117 per month. Of these, 303 were classified as significant incidents, averaging around 25 per month. Throughout the same year, ACN's specialist staff intervened on-site in 13 cases and provided remote support in 31 cases. Additionally, the NCS convened 11 times in its regular format and 7 times in a restricted format. The CVCN identified 38 zero-day vulnerabilities, 22 of which were categorized as "high-level" or "critical" (ACN Annual Report, 2024). Geopolitical tensions due to the Russo-Ukrainian conflict and the events post-October 7 in Israel have heightened global threats, impacting Italy as well. Furthermore, the risk index is expected to rise with the increasing malicious usage of AI technologies for economic, political, and military purposes. The expansion of Internet usage (i.e. the Internet of Everything) and AI will broaden the attack surface, making society more vulnerable to coordinated malicious actions. This issue is exacerbated by the fact that states are also leveraging these

L. Martino, *Cybersecurity in Italy*, SpringerBriefs in Cybersecurity,
https://doi.org/10.1007/978-3-031-64396-5_7

technologies for political and military ends. As Deibert notes, we are experiencing "growing pressures on governments and their armed forces to develop the capacity to fight and win wars in this domain" (Deibert, 2011, p. 2). Therefore, it is not only individual and private security that is at risk, but also collective and national interests. In other words, without effective national cybersecurity governance, the foundational assets of an entire social system are at risk (Dunn Cavelty, 2024). In order to manage that risks, over the past decade, Italy has undergone a profound transformation, marked by the implementation of three significant institutional reforms that have shaped the national approach to cybersecurity. The Italian cybersecurity policy framework has evolved through various phases, reflecting a dynamic interplay among different actors and stakeholders. Beginning with the "Monti Decree" of 2013, which placed cybersecurity under the direct responsibility of the Prime Minister, the need for a revision of existing policies became evident due to their shortcomings. This decree introduced the first form of Italian architectural governance for cybersecurity. However, despite efforts to improve the national cybersecurity framework, several major deficiencies soon emerged, including overlaps between competent authorities, dispersion of responsibilities, lack of transparency in decision-making, diversity of involved actors, and the absence of a specific budget to achieve strategic objectives. These issues, compounded by European and NATO initiatives, highlighted the necessity for reform. The "Renzi Directive" of 2015 and the subsequent "Gentiloni Decree" of 2017 sought to address these problems by reorganizing the national architecture. The second reform improved the distribution of responsibilities and decision-making processes, while also incorporating European-derived recommendations and regulations. Nonetheless, fundamental questions about governance efficiency and the effectiveness of implemented policies persisted. One such question was whether Intelligence services were the most appropriate public administration apparatus for managing cybersecurity, given that transparency and direct interactions among stakeholders are crucial. The approval of the Perimeter Law (PSNC) not only marked a strategic turning point in Italian cybersecurity legislation, but also intensified these questions. Specifically, decision-makers began to ask: Can intelligence services, which are inherently secretive, effectively manage cybersecurity where information sharing is critical? In response to this question, Italy decided in 2021 to evolve the policy framework by approving a third reform. This reform revolutionized the national cybersecurity governance and led to the creation of the National Cybersecurity Agency. The change reframed cybersecurity, as not merely a domain reserved for intelligence services, defense, or law enforcement, but as a critical element for ensuring the political, social, economic, and industrial resilience of the nation. However, several important open questions remain. For example, will centralizing both oversight and coordination functions within a single Agency be an effective and efficient long-term governance system? What are the agency's actual capabilities in dealing with the expanding threat landscape presented by artificial intelligence (AI) and disruptive technologies like quantum computing? How well can the Agency intervene across the entire national territory? What is the most effective model for involving private actors, who are increasingly burdened by stringent regulatory requirements? Furthermore, what best practices could be adopted in Italy to enhance the current governance? If these questions are not systematically addressed, there is a significant risk of undermining the stability of the entire national ecosystem.

References

Cavelty. 2024. *Myriam Dunn Cavelty, The Politics of Cyber-Security*. Routledge.
Deibert. 2011. Ronald Deibert: Tracking the emerging arms race in cyberspace. An Interview. *Bulletin of the Atomic Scientists* 67 (2): 1–8.

List of Concept and Acronyms

Acronyms and concept	Description
ACN	National Cybersecurity Agency
AISE	External Intelligence and Security Agency
AISI	Internal Intelligence and Security Agency
Authentication	Verifying the identity of users and systems to ensure that only authorized entities have access to resources
Authorization	Granting appropriate permissions to users and systems based on their roles and responsibilities
Availability	Being able to ensure that systems and data are available and accessible when needed
CERT	Computer Emergency Response Team
CeVa	Evaluation Center
CIC	Interministerial Committee for Cybersecurity
CINI	National Interuniversity Consortium for Informatics
CISR	Interministerial Committee for the Security of the Republic
CNAIPIC	National Anti-Crime Computer Center for the Protection of Critical Infrastructures
Confidentiality	Guaranteeing that data remains private and accessible only to authorized users
Coordination and information sharing	Collaborate with government agencies, private sector organizations, and international partners to share information on threats and respond effectively to cyber incidents
COPASIR	Parliamentary Committee for the Security of the Republic
COR	Network Operations Command
CSIRT	Computer Security Incident Response Team
CVCN	National Assessment and Certification Center
Cyber diplomacy	Having a diplomatic apparatus capable of engaging in diplomatic efforts to establish norms, rules, and agreements in the international community to reduce the risk of cyber conflicts and implement multilateral and bilateral collaborations

L. Martino, *Cybersecurity in Italy*, SpringerBriefs in Cybersecurity, https://doi.org/10.1007/978-3-031-64396-5

Acronyms and concept	Description
Cyber economic security	The capacity to prevent cyber espionage and the theft of intellectual property that could undermine a nation's economic competitiveness
Cyber espionage and counterespionage	Monitor and counter foreign cyber espionage and intelligence activities targeting a nation's government, industry, and public and research institutions
Cyber national defense	The need to protect military systems and communications and implement tools useful for national defense against cyberattacks that could compromise national security and social stability
Deterrence and response	Implement suitable strategies to deter cyber threats through a combination of defensive and offensive measures, international cooperation, and the ability to respond effectively to cyberattacks when they occur
DIS	Department of Information for Security
DPCM	Decree of the President of the Council of Ministers
EU	European Union
ICT	Information and communication technologies
Integrity	Ensuring the accuracy and reliability of data and systems
Intrusion detection and prevention	Implementing capabilities to monitor unauthorized access or malicious activity and taking measures to prevent or mitigate such situations
IT	Information technologies
NATO	North Atlantic Treaty Organization
Network security	Implementing network security measures, intrusion detection systems, and intrusion prevention systems to protect against external and internal threats
NISP	Interministerial Situation and Planning Unit
NSC	Cybersecurity Unit
OES	Operators of essential services
OSCE	Organization for Security and Co-operation in Europe
Protection of critical infrastructure	The ability to safeguard essential and vital sectors for the social and economic functioning of a state, such as energy, transportation, finance, and healthcare, from cyber threats to ensure security and stability
PSNC	National cybersecurity perimeter
Public security	Ensure the safety and well-being of citizens by protecting critical public services, such as emergency response systems and healthcare infrastructure, from cyber threats
Security updates and patches	Enforcing internal policies regularly, such as security patches and updates to software and computer systems, to address known vulnerabilities and ensure a continuous flow of vulnerability awareness
Vulnerability management	Identifying and addressing security vulnerabilities in software, hardware, and configurations to reduce the attack surface

Further Reading

Anders, Therese, Christopher J. Fariss, and Jonathan N. Markowitz. 2019. Bread before Guns or Butter: Introducing Surplus Domestic Product (SDP). *International Studies Quarterly*.

Anderson, R.H., and Hearn, A.C. 1996. An Exploration of Cyberspace Security R&D Investment Strategies for DARPA: "The Day After . . . in Cyberspace II." RAND Corporation.

Arquilla, J., and D. Ronfeldt, eds. 1997. *In Athena's Camp: Preparing for Conflict in the Information Age*. RAND Corporation.

Baldwin, D.A. 1997. The Concept of Security. *Review of International Studies* 23 (1): 5–26.

Baldwin, David A. 2012. Power and International Relations. In *Handbook of International Relations*, 273.

Baliga, S., E. Bueno de Mesquita, and A. Wolitzky. 2020. Deterrence with Imperfect Attribution. *American Political Science Review* 114 (4): 1155–1178.

Balzacq, T., D. Basaran, D. Bigo, E. Guittet, and C. Olsson. 2010. Security Practices. In: *Oxford Research Encyclopedia of International Studies*. https://oxfordre.com/internationalstudies/view/10.1093/acrefore/9780190846626.001.0001/ac refore-9780190846626-e-475

Barbrook, R., and A. Cameron. 1995. The Californian Ideology. *Mute* 1 (3) www.metamute.org/editorial/articles/californian-ideology.

Bayuk, J., J. Healey, P. Rohmeyer, M.H. Sachs, J. Schmidt, and J. Weiss. 2012. *Cyber Security Policy Guidebook*. Wiley.

Beckley, Michael. 2018. The Power of Nations: Measuring What Matters. *International Security* 43 (2): 7–44.

Bendrath, R. 2001. The Cyberwar Debate: Perception and Politics in US Critical Infrastructure Protection. *Information & Security: An International Journal* 7: 80–103.

Bequai, A. 1986. *Technocrimes: The Computerization of Crime and Terrorism*. Lexington Books.

Betz, D., and T. Stevens. 2011. *Cyberspace and the State: Toward a Strategy for Cyber- Power*. Adelphi Papers, vol. 424. The International Institute for Strategic Studies.

Betz, David, and Tim Stevens. 2017. *Cyberspace and the State: Toward a Strategy for Cyber-Power*.

Bingham, N. 1996. Objections: From Technological Determinism towards Geographies of Relations. *Environment and Planning D: Society and Space* 14 (6): 635–657.

Bossong, R., and B. Wagner. 2017. A Typology of Cyber-Security and Public-Private Partnerships in the Context of the EU. *Crime, Law and Social Change. An International Journal* 67 (3): 265–288.

Branch, J. 2021. What's in a Name? *Metaphors and Cybersecurity. International Organization* 75 (1): 39–70.

Broeders, D., L. Adamson, and R. Creemers. 2019. Coalition of the Unwilling? Chinese and Russian Perspectives on Cyberspace. In *The Hague Program for Cyber Norms Policy Brief*, November. https://ssrn.com/abstract=3493600

Brown, K.A. 2006. *Critical Path: A Brief History of Critical Infrastructure Protection in the United States*. George Mason University Press.

Brown, A.J. 2020. "Should I Stay or Should I Leave?": Exploring (Dis)continued Facebook Use After the Cambridge Analytica Scandal. *Social Media + Society* 6 (1). https://doi.org/10.1177/2056305120913884.

Brunner, C. 2021. Conceptualizing Epistemic Violence: An Interdisciplinary Assemblage for IR. *International Political Review* 9 (1): 193–212.

Buchanan, B. 2016. *The Cybersecurity Dilemma: Hacking, Trust, and Fear Between Nations*.

———. 2020. *The Hacker and the State: Cyber Attacks and the New Normal of Geopolitics*. Harvard University Press.

Buzan, B., and L. Hansen. 2009. *The Evolution of International Security Studies*. Cambridge University Press.

Buzan, B., O. Wæver, and J. de Wilde. 1998. *Security: A New Framework for Analysis*. Lynne Rienner.

C.A.S.E. Collective. 2006. Critical Approaches to Security in Europe: A Networked Manifesto. *Security Dialogue* 37 (4): 443–487.

Carr, M. 2016. Public-Private Partnerships in National Cyber-Security Strategies. *International Affairs* 92 (1): 43–62.

Cerny, Philip G. 2010. *Rethinking World Politics: A Theory of Transnational Neopluralism*. Oxford: Oxford University Press.

Chesney, R., and M. Smeets, eds. 2023. *Deter, Disrupt, or Deceive: Assessing Cyber Conflict as an Intelligence Contest*. Georgetown University Press.

Choucri, N., and D.D. Clark. 2019. *International Relations in the Cyber Age: The Co- evolution Dilemma*. MIT Press.

Christensen, K.K., and K.L. Petersen. 2017. Public–Private Partnerships on Cyber Security: A Practice of Loyalty. *International Affairs* 93 (6): 1435–1452.

Clarke, R. 2016. Big Data. *Big Risks. Information Systems Journal* 26 (1): 77–90.

Cohen, J.E. 2007. Cyberspace as/and Space. *Columbia Law Review* 107 (1): 210–256.

Collier, S.J., and A. Lakoff. 2008. How Infrastructure Became a Security Problem. In *The Politics of Securing the Homeland: Critical Infrastructure, Risk, and Securitisation*, ed. M. Dunn and K.S. Kristensen, 17–39. Routledge.

Collins, A. 2019. *Contemporary Security Studies*. Oxford University Press.

Columba, P., and N. Vaughan-William. 2010. *Critical Security Studies: An Introduction*. Routledge.

Computer Science and Telecommunications Board. 1989. *Growing Vulnerability of the Public Switched Network: Implications for National Security Emergency Preparedness*. National Academy Press.

Cristiano, F., X. Kurowska, T. Stevens, L.M. Hurel, N.S. Fouad, M. Dunn Cavelty, D. Broeders, T. Liebetrau, and J. Shires. 2024. Cybersecurity and the Politics of Knowledge Production: Towards a Reflexive Practice. *Journal of Cyber Policy*. https://doi.org/10.1080/23738871.2023.2287687.

Cyjax. 2022. *Who is Trickbot? Analysis of the Trickbot Leaks*. www.cyjax.com/wp-content/uploads/2022/07/Who-is-Trickbot.pdf

Dahl, Robert A. 1957. The Concept of Power. *Behavioral Science* 2 (3): 201–215.

Dahl, R.A. 1963. *Modern Political Analysis*. Prentice-Hall.

de Leeuw, K., and J. Bergstra. 2007. *The History of Information Security: A Comprehensive Handbook*. Elsevier Science.

Dean, B., and R. McDermott. 2017. A Research Agenda to Improve Decision Making in Cyber Security Policy. *Penn State Journal of Law & International Affairs* 5 (1): 29–164.

Deibert, R. 2018. Toward a Human-Centric Approach to Cyber-Security. *Ethics & International Affairs* 32 (4): 411–424.

Demchak, C.C. 2011. *Wars of Disruption and Resilience: Cybered Conflict, Power, and National Security*. University of Georgia Press http://www.jstor.org/stable/j.ctt46np69.

Derian, J.D. 2000. Virtuous War/Virtual Theory. *International Affairs* 76 (4): 771–788.

Diffie, W., and M.E. Hellman. 1976. New Directions in Cryptography. *IEEE Transactions on Information Theory* 22 (6): 644–654.

Dunn, M. 2002. *Information Age Conflicts: A Study on the Information Revolution and a Changing Operating Environment*. Zürcher Beiträge zur Sicherheit-spolitik und Konfliktforschung, No. 64. Center for Security Studies.

Dunn Cavelty, M. 2008. *Cyber-Security and Threat Politics: US Efforts to Secure the Information Age*. Routledge.

———. 2014. Breaking the Cyber-Security Dilemma: Aligning Security Needs and Removing Vulnerabilities. *Science and Engineering Ethics* 20 (3): 701–715.

———. 2019. The Materiality of Cyberthreats: Securitization Logics in Popular Visual Culture. *Critical Studies on Security* 7 (2): 138–151.

———. 2020. Cyber-Security Between Hypersecuritization and Technological Routine. In *Routledge Handbook of International Cyber- security*, ed. E. Tikk and M. Kerttunen, 11–21. Routledge.

Dunn Cavelty, M., and T. Balzacq. 2017. *The Routledge Handbook of Security Studies*. Routledge.

Dunn Cavelty, M., and F. Egloff. 2019a. The Politics of Cyber-security: Balancing Different Roles of the State. *St Antony's International Review* 5 (1): 37–57.

Dunn Cavelty, M., and F.J. Egloff. 2019b. The Politics of Cyber-Security: Balancing Different Roles of the State. *St Antony's International Review* 15 (1): 37–57.

Dunn Cavelty, M., C. Eriksen, and B. Scharte. 2023. Making Cyber Security More Resilient: Adding Social Considerations to Technological Fixes. *Journal of Risk Research* 26 (7): 801–814.

Dunn Cavelty, M., and J. Hagmann. 2021. The Politics of Security and Technology in Switzerland. *Swiss Political Science Review* 27 (1): 128–138.

Dunn Cavelty, M., M. Kaufmann, and S.K. Kristensen. 2015. Resilience and (In) security: Practices, Subjects. *Temporalities. Security Dialogue* 46 (1): 3–14.

Dunn Cavelty, M., T. Pulver, and M. Smeets. 2024. The Evolution of Cyber Conflict Studies. *International Affairs*.

Dunn Cavelty, M., and M. Suter. 2009. Public-Private Partnerships Are No Silver Bullet: An Expanded Governance Model for Critical Infrastructure Protection. *International Journal of Critical Infrastructure Protection* 4 (2): 179–187.

Dunn Cavelty, M., and A. Wenger. 2020. Cyber Security Meets Security Politics: Complex Technology, Fragmented Politics, and Networked Science. *Contemporary Security Policy* 41 (1): 5–32.

Dunn, M., and S.K. Kristensen, eds. 2008. *The Politics of Securing the Homeland: Critical Infrastructure, Risk, and Securitisation*. Routledge.

Dunne, J. Paul, and Ron P. Smith. 2019. Military Expenditure, Investment and Growth. *Defence and Peace Economics*: 1–14.

Dwyer, A. 2023. Cybersecurity's Grammars: A More-than-human Geopolitics of Computation. *Area* 55 (1): 10–17.

Dwyer, A., C. Stevens, L.P. Muller, M. Dunn Cavelty, L. Coles-Kemp, and P. Thornton. 2022. What Can a Critical Cybersecurity Do? *International Political Sociology* 16 (3): olac013. https://doi.org/10.1093/ips/olac013.

Easton, D. 1965. *A Framework for Political Analysis*. Prentice-Hall.

Edwards, P.N. 1997. *The Closed World Computers and the Politics of Discourse in Cold War America*. MIT Press.

Egloff, F.J. 2020. Public Attribution of Cyber Intrusions. *Journal of Cybersecurity* 6 (1): tyaa012.

Fayi, S.Y.A. 2018. What Petya/NotPetya Ransomware Is and What Its Remidiations Are. In *Information Technology – New Generations. 15th International Conference on Information Technology*, ed. S. Latifi, 93–100. Springer.

Federation of American Scientists. 1996, June 5. *Security in Cyberspace: U.S. Senate Permanent Subcommittee on Investigations* (Minority Staff Statement). Appendix B. The Case Study: Rome Laboratory, Griffiss Air Force Base, NY Intrusion. https://irp.fas.org/congress/1996_hr/s960605b.htm

Flynn, K. 2006. Covert Disclosures: Unauthorized Leaking, Public Officials and the Public Sphere. *Journalism Studies* 7 (2): 256–273.

Fritsch, S. 2014. Conceptualizing the Ambivalent Role of Technology in International Relations: Between Systemic Change and Continuity. In *The Global Politics of Science and Technology – Vol. 1: Concepts From International Relations and Other Disciplines*, ed. M. Mayer, M. Carpes, and R. Knoblich, 115–138. Springer.

Galbraith, J.K. 1985. *The New Industrial State*. Princeton University Press.

Gallie, W.B. 1956. Essentially Contested Concepts. *Proceedings of the Aristotelian Society* 56 (1): 167–198.

Georgieva, I. 2020. The Unexpected Norm-Setters: Intelligence Agencies in Cyberspace. *Contemporary Security Policy* 41 (1): 33–54.

Giacomello, Giampiero, Iovanella Antonio, and Luigi Martino. 2023. A Small World of Bad Guys: Investigating the Behavior of Hacker Groups in Cyber-Attacks. ArXiv Pre-print. https://arxiv.org/abs/2309.16442

Gibson, W. 1989. *Neuromancer*. Berkley Publishing.

Giles, K., and W. Hagestad. 2013. Divided by a Common Language: Cyber-Definitions in Chinese, Russian and English. In *Proceedings of the 5th International Conference on Cyber-Conflict*, ed. K. Podins, J. Stinissen, and M. Maybaum, 1–17. CCD COE Publications.

Gilpin, Robert. 1987. *The Political Economy of International Relations*. Princeton: Princeton University Press.

Goldstone, Jack A., Eric P. Kaufmann, and Monica Duffy Toft. 2012. *Political Demography: How Population Changes Are Reshaping International Security and National Politics*. Oxford: Oxford University Press.

Gomez, M.A. 2019. Sound the Alarm! Updating Beliefs and Degradative Cyber-Operations. *European Journal of International Security* 4 (2): 190–208.

Gomez, M.A., and E.B. Villar. 2018. Fear, Uncertainty, and Dread: Cognitive Heuristics and Cyber Threats. *Politics and Governance* 6 (2): 61–72.

Gorwa, Robert, and Max Smeets. 2019a. *Cyber Conflict in Political Science: A Review of Methods and Literature*. https://doi.org/10.31235/osf.io/fc6sg.

Gorwa, R., and M. Smeets. 2019b, July 25. *Cyber Conflict in Political Science: A Review of Methods and Literature*. SocArXiv Papers. https://doi.org/10.31235/osf.io/fc6sg

Graham, S. 1998. The End of Geography or the Explosion of Place? Conceptualizing Space, Place and Information Technology. *Progress in Human Geography* 22 (2): 165–185.

Greenberg, A. 2018. The Untold Story of NotPetya, the Most Devastating Cyber-attack in History. *Wired*, August 22. https://www.wired.com/story/notpetya-cyberattack- ukraine-russia-code-crashed-the-world/

———. 2019. *Sandworm: A New Era of Cyberwar and the Hunt for the Kremlin's Most Dangerous Hackers*. Doubleday.

Guerrero-Saade, J.A., C. Raiu, D. Moore, and T. Rid. 2018. *Penquin's Moonlit Maze: The Dawn of Nation-State Digital Espionage*. https://media.kasperskycontenthub.com/wp-content/uploads/sites/43/2018/03/07180251/Penquins_Moonlit_Maze_PDF_eng.pdf

Gygli, Savina, Florian Haelg, Niklas Potrafke, and Jan-Egbert Sturm. 2019. The KOF Globalisation Index–Revisited. *The Review of International Organizations* 14 (3): 543–574.

Hagmann, J. 2015. *(In-)Security and the Production of International Relations*. Routledge.

Harris, K. 2023. Epistemic Domination. Thought: A. *Journal of Philosophy*. https://doi.org/10.5840/tht202341317.

Healey, J., ed. 2013. *A Fierce Domain: Conflict in Cyberspace, 1986 to 2012*. Cyber Conflict Studies Association.

Herr, T., A.P.B. Laudrain, and M. Smeets. 2020. Mapping the Known Unknowns of Cyber- Security Education: A Review of Syllabi on Cyber-Conflict and Security. *Journal of Political Science Education* 17 (Supp. 1): 503–519.

Herrera, G.L. 2003. Technology and International Systems. *Millennium Journal of International Studies* 32 (3): 559–593.

Hill, K. 2013. Thanks, Snowden! Now All the Major Tech Companies Reveal How Often They Give Data to Government. *Forbes*, November 14. www.forbes.com/sites/kashmirhill/2013/11/14/silicon-valley-data-handover-infographic/?sh=7f6a67945365

Hoogensen, G., and K. Stuvøy. 2006. Gender, Resistance and Human Security. *Security Dialogue* 37 (2): 207–228.

Hurel, L.M., and L.C. Lobato. 2018. Unpacking Cyber-Norms: Private Companies as Norm Entrepreneurs. *Journal of Cyber-Policy* 3 (1): 61–67.

Huysmans, J. 2000. The European Union and the Securitization of Migration. *Journal of Common Market Studies* 38 (5): 751–777.

———. 2006. *The Politics of Insecurity. Fear, Migration and Asylum in the EU*. Routledge.

Ignazi, Piero, Giampiero Giacomello, and Fabrizio Coticchia. 2012. *Italian Military Operations Abroad: Just Don't Call It War*. London: Palgrave Macmillan.

International Institute for Strategic Studies. 2015. *Evolution of the Cyber-Domain: The Implications for National and Global Security*. Routledge.

———. 2021. *Cyber Capabilities and National Power: A Net Assessment*. www.iiss.org/research-paper//2021/06/cyber-capabilities- national-power

Jackson, P.T., and D.H. Nexon. 1999. Relations Before States: Substance, Process and the Study of World Politics. *European Journal of International Relations* 5 (3): 291–332.

Jasanoff, S., ed. 2004. *States of Knowledge: The Co-production of Science and Social Order*. Routledge.

Kello, L. 2013. The Meaning of the Cyber Revolution: Perils to Theory and Statecraft. *International Security* 38 (2): 7–40.

Keohane, Robert Owen, and Joseph S. Nye. 1977. *Power and Interdependence: World Politics in Transition*. Boston: Little, Brown.

Khokhlov, N., and A. Korotayev. 2022. Internet, Political Regime and Terrorism: A Quantitative Analysis. *Cross-Cultural Research* 56 (4): 385–418. https://doi.org/10.1177/10693971221085343.

Kingdon, J.W. 1984. *Agendas, Alternatives, and Public Policies*. Harper Collins College.

Kissel, R. 2013. Glossary of Key Information Security Terms. *NIST Interagency Reports NIST IR* 7298 (3). https://doi.org/10.6028/NIST.IR.7298r3.

Kostyuk, N. 2020. *Public Cyberinstitutions: Signaling State Cybercapacity*. PhD Thesis University of Michigan. https://deepblue.lib.umich.edu/bit-stream/handle/2027.42/163168/nadiya_1.pdf?sequence=1

Kramer, F.D., S.H. Starr, and L. Wentz, eds. 2009. *Cyberpower and National Security*. NDU Press: Center for Technology and National Security Policy.

Krause, K. 2019. Technologies of Violence. Myriam Dunn Cavelty and Jonas Hagmann in Conversation With Keith Krause. In *Technologies of International Relations: Continuity and Change*, ed. C. Kaltofen, M. Carr, and M. Acuto, 97–106. Palgrave Macmillan.

Kristensen, K.S. 2008. "The Absolute Protection of our Citizens": Critical Infrastructure Protection and the Practice of Security. In *The Politics of Securing the Homeland: Critical Infrastructure, Risk and Securitisation*, ed. M. Dunn and K.S. Kristensen, 63–83. Routledge.

Kruck, A., and M. Weiss. 2023. The Regulatory Security State in Europe. *Journal of European Public Policy* 30 (7): 1205–1229.

Leal, M.M., and P. Musgrave. 2023. Backwards from Zero: How the U.S. Public Evaluates the Use of Zero-Day Vulnerabilities in Cybersecurity. *Contemporary Security Policy* 44 (3): 437–461.

Leese, M., and M. Hoijtink, eds. 2019. *Technology and Agency in International Relations*. Routledge.

Leiner, B.M., V.G. Cerf, D.D. Clark, R.R. Kahn, L. Kleinrock, D.C. Lynch, J. Postel, L.G. Roberts, and S. Wolff. 2009. A Brief History of the Internet. *ACM SIGCOMM Computer Communication Review* 39 (5): 22–31.

Lessig, L. 1999. *Code and Other Laws of Cyberspace*. Basic Books.

Libicki, M. 2000. *Who Runs What in the Global Information Grid: Ways to Share Local and Global Responsibility*. RAND Corporation.

Liebetrau, T., and K.K. Christensen. 2021. The Ontological Politics of Cyber-Security: Emerging Agencies, Actors, Sites and Spaces. *European Journal of International Security* 6 (1): 25–43.

Lilli, E. 2023. How Can We Know What We Think We Know About Cyber Operations? *Journal of Global Security Studies* 8 (2): 1–18. https://doi.org/10.1093/jogss/ogad011.

Lindsay, J.R. 2015. Tipping the Scales: The Attribution Problem and the Feasibility of Deterrence Against Cyberattack. *Journal of Cyber-Security* 1 (1): 53–67.

———. 2017. Restrained by Design: The Political Economy of Cyber-Security. *Digital Policy, Regulation and Governance* 19 (6): 493–514.

Lupovici, A. 2016. The "Attribution Problem" and the Social Construction of "Violence": Taking Cyber-Deterrence Literature a Step Forward. *International Studies Perspectives* 17 (3): 322–342.

Macnish, K. 2021. An End to Encryption? Surveillance and Proportionality in the Crypto- Wars. In *Counter-Terrorism, Ethics and Technology: Advanced Sciences and Technologies for Security Applications*, ed. A. Henschke, A. Reed, S. Robbins, and S. Miller. Springer.

Martelle, M. (ed.). 2018. *Eligible Receiver 97: Seminal DOD Cyber Exercise Included Mock Terror Strikes and Hostage Simulations*. The Cyber Vault Project, Briefing Book #634. https://nsarchive.gwu.edu/briefing-book/cyber-vault/2018-08-01/eligible-receiver-97-seminal-dod-cyber-exercise-included-mock-terror-strikes-hostage-simulations

Martino Luigi, and Federica Merenda .2021. Artificial Intelligence: A Paradigm Shift in International Law and Politics? Autonomous Weapon Systems as a Case Study. In *Technology and International Relations The New Frontier in Global Power*, ed. Giampiero Giacomello, Francesco Moro, Marco Valigi. Edward Elgar. ISBN: 9781788976060.

Maschmeyer, L., R.J. Deibert, and J.R. Lindsay. 2021. A Tale of Two Cybers How Threat Reporting by Cybersecurity Firms Systematically Underrepresents Threats to Civil Society. *Journal of Information Technology & Politics* 18 (1): 1–20.

Matania, E., and U. Sommer. 2023. Tech Titans, Cyber Commons and the War in Ukraine: An Incipient Shift in International Relations. *International Relations*. https://doi.org/10.1177/00471178231211500.

Matthewman, S. 2011. *Technology and Social Theory*. Palgrave Macmillan.

May, C., M. Baker, D. Gabbard, T. Good, G. Grimes, M. Holmgren, et al. 2018. *Advanced Information Assurance Handbook*. Report: Carnegie Mellon University. https://doi.org/10.1184/R1/6571844.v1.

Mayer, M., M. Carpes, and R. Knoblich. 2014. The Global Politics of Science and Technology: An Introduction. In *The Global Politics of Science and Technology*, ed. M. Mayer, M. Carpes, and R. Knoblich, 1–35. Springer.

McCarthy, D.R. 2018. Introduction: Technology in World Politics. In *Technology and World Politics: An Introduction*, ed. D.R. McCarthy, 1–21. Routledge.

Meiske, Maline. 2019. *Burden-Sharing in Peace Operations: Quantitative Studies on the Global, Regional, and National Level*. Oxford: University of Oxford.

Middleton, B. 2017. *A History of Cyber Security Attacks: 1980 to Present*. Auerbach Publications.

Molander, R.C., A.S. Riddle, and P.A. Wilson. 1996. *Strategic Information Warfare: A New Face of War*. RAND Corporation.

Morgenthau, Hans Joachim. 1948. *Politics among nations*. New York.

Muller, L.P., and N. Welfens. 2023. (Not) Accessing the Castle: Grappling With Secrecy in Research on Security Practices. *Secrecy and Society* 3 (1). https://doi.org/10.55917/2377-6188.1073.

Mumford, D., and J. Shires. 2023. Toward a Decolonial Cybersecurity: Interrogating the Racial-Epistemic Hierarchies That Constitute Cybersecurity Expertise. *Security Studies* 32 (4–5): 622–652.

Mungo, P., and B. Clough. 1992. *Approaching Zero: The Extraordinary Underworld of Hackers, Phreakers, Virus Writers, and Keyboard Criminals*. Random House.

Murphy, C.C. 2020. The Crypto-Wars Myth: The Reality of State Access to Encrypted Communications. *Common Law World Review* 49 (3–4): 245–261.

National Cyber Strategy of the United Kingdom https://www.gov.uk/government/publications/national-cyber-strategy-2022/national-cyber-security-strategy-2022

Naughton, J. 2016. The Evolution of the Internet: From Military Experiment to General Purpose Technology. *Journal of Cyber Policy* 1 (1): 5–28.

Neal, A.W. 2019. *Security as Politics: Beyond the State of Exception*. Edinburgh University Press.

Nelkin, D. 1992. *Controversy: Politics of Technical Decisions*. 3rd ed. Sage.

Nissenbaum, H. 2005. Where Computer Security Meets National Security. *Ethics and Information Technology* 7 (2): 61–73.

Norman, A.R.D. 1983. *Computer Insecurity*. Chapman and Hall.

Nye, Joseph S. Jr. 2010. *Cyber Power*. Belfer Center for Science and International Affairs. Harvard Kennedy School.

Ould Mohamedou, M.M. 2020. In Search of the Non-Western State: Historicising and De-Westphalianising Statehood. In *The SAGE Handbook of Political Science*, ed. D. Berg-Schlosser, B. Badie, and L. Morlino, 1335–1348. Sage.

Parikka, J. 2005. *Digital Contagions: A Media Archaeology of Computer Viruses*. Peter Lang.

Parker, D.B. 1983. *Fighting Computer Crime*. Charles Scribner's Sons.

Perelman, L.J. 2007. Shifting Security Paradigms: Toward Resilience. In *Critical Thinking: Moving from Infrastructure Protection to Infrastructure Resilience*, CIP Program Discussion Paper Series, ed. J.A. McCarthy, 23–48. George Mason University.

Petersen, K.L. 2012. Risk Analysis – A Field Within Security Studies? *European Journal of International Relations* 18 (4): 693–717.

Petersen, K.L., and V.S. Tjalve. 2013. (Neo)Republican Security Governance? US Homeland Security and the Politics of "Shared Responsibility". *International Political Sociology* 7 (1): 1–18.

Powers, S.M., and M. Jablonski. 2015. *The Real Cyber War: The Political Economy of Internet Freedom*. University of Illinois Press.

President's Commission on Critical Infrastructure Protection. 1997. *Critical Foundations: Protecting America's Infrastructures*. US Government Printing Office.

Price, M. 2018. The Global Politics of Internet Governance: A Case Study in Closure and Technological Design. In *Technology and World Politics: An Introduction*, ed. D.R. McCarthy, 126–145. Routledge.

Rattray, G. 2001. *Strategic Warfare in Cyberspace*. MIT Press.

Reagan, R. 1984. *National Policy on Telecommunications and Automated Information Systems Security*. National Security Decision Directive NSDD 145, September 17. The White House.

Rid, T., and B. Buchanan. 2015. Attributing Cyber Attacks. *Journal of Strategic Studies* 38 (1–2): 4–37.

Romaniuk, S.N., and M. Manjikian, eds. 2021. *Routledge Companion to Global Cyber- Security Strategy*. Routledge.

Rosenau, James N. 2018. *Turbulence in World Politics: A Theory of Change and Continuity*. Princeton University Press.

Ross, A. 1990. Hacking Away at the Counterculture. *Postmodern Culture* 1 (1). https://doi.org/10.1353/pmc.1990.0011.

Sabillon, R., V. Cavaller, and J. Cano. 2016. National Cyber-Security Strategies: Global Trends in Cyberspace. *International Journal of Computer Science and Software Engineering* 5 (5): 67–81.

Scherlis, W.L., S.L. Squires, and R.D. Pethia. 1990. Computer Emergency Response. In *Computers Under Attack: Intruders, Worms, and Viruses*, ed. P. Denning, 495–504. Addison-Wesley.

Schulze, M. 2017. Clipper Meets Apple vs. FBI – A Comparison of the Cryptography Discourses from 1993 and 2016. *Media and Communication* 5 (1): 54–62.

Schwartau, W. 1994. *Information Warfare: Chaos on the Electronic Super Highway.* Thunder's Mouth.

Seifert, F., and C. Fautz. 2021. Hype after Hype: From Bio to Nano to AI. *Nanoethics* 15 (2): 143–148.

Shafqat, N., and A. Masood. 2016. Comparative Analysis of Various National Cyber Security Strategies. *International Journal of Computer Science and Information Security* 14 (1): 129–136.

Shandler, R., and M.A. Gomez. 2023. The Hidden Threat of Cyber-Attacks – Undermining Public Confidence in Government. *Journal of Information Technology & Politics* 20 (4): 359–374.

Shandler, R., M.L. Gross, S. Backhaus, and D. Canetti. 2022. Cyber-Terrorism and Public Support for Retaliation–A Multi-Country Survey Experiment. *British Journal of Political Science* 52 (2): 850–868.

Sharma, R.K. 2022. James Shires (2021). The Politics of Cybersecurity in the Middle East. Hurst & Co. Hardback. ISBN: 9781787384736. 361 pp. *Contemporary Review of the Middle East* 9 (3): 374–376. https://doi.org/10.1177/23477989221099981.

Sharma, A., and A. Gupta, eds. 2006. *The Anthropology of the State: A Reader.* Blackwell.

Sieber, U. 1986. *The International Handbook on Computer Crime: Computer-Related Economic Crime and the Infringements of Privacy.* Wiley.

Sismondo, S. 2010. *An Introduction to Science and Technology Studies.* 2nd ed. Blackwell.

Skibell, R. 2002. The Myth of the Computer Hacker. *Information, Communication & Society* 5 (3): 336–356.

Smeets, M. 2022. A US History of not Conducting Cyber Attacks. *Bulletin of the Atomic Scientists* 78 (4): 208–213.

Snider, K., S. Zandani, R. Shandler, and D. Canetti. 2021. Cyber-Terrorism, Cyber- Threats and Attitudes Toward Cyber-Security Policies. *Journal of Cyber-security* 7 (1): tyab019. https://doi.org/10.1093/cybsec/tyab019.

Soesanto, S. 2019. *The Evolution of US Defense Strategy in Cyberspace (1988–2019).* CSS Cyber Defense Trend Analysis. https://css.ethz.ch/content/dam/ethz/special-interest/gess/cis/center-for-securities-studies/pdfs/Cyber-Reports-2019-08-The-Evolution-of-US-defense-strategy-in-cyberspace.pdf

———. 2023. Ukraine's IT Army. *Survival* 65 (3): 93–106.

Solar, C. 2020. Cyber-Security and Cyber Defence in the Emerging Democracies. *Journal of Cyber Policy* 5 (3): 392–412.

Spafford, E.H. 1989. The Internet Worm: Crisis and Aftermath. *Communications of the ACM* 32 (6): 678–687.

Srinivas, J., A. Kumar Das, and N. Kumar. 2019. Government Regulations in Cyber- Security: Framework, Standards and Recommendations. *Future Generation Computer Systems* 92: 178–188.

Stevens, T. 2018. Global Cybersecurity: New Directions in Theory and Methods. *Politics and Governance* 6 (2). https://doi.org/10.17645/pag.v6i2.1569.

Stevens, C. 2020. Assembling Cyber-security: The Politics and Materiality of Technical Malware Reports and the Case of Stuxnet. *Contemporary Security Policy* 41 (1): 129–152.

Stevens, T. 2023. *What is Cybersecurity for?* Bristol University Press.

Stoll, C. 1989. *The Cuckoo's Egg: Tracking a Spy through the Maze of Computer Espionage.* Doubleday.

The White House. 2023. *National Cybersecurity Strategy.* The White House. www.whitehouse.gov/wp-content/uploads/2023/03/National-Cybersecurity-Strategy-2023.pdf

Tilly, C. 1985. War Making and State Making as Organized Crime. In *Bringing the State Back in*, ed. P. Evans, D. Rueschemeyer, and T. Skocpol, 169–191. Cambridge University Press.

Tropina, T., and C. Callanan. 2015. *Self- and Co-regulation in Cybercrime, Cybersecurity and National Security*, Springer Briefs in Cybersecurity. Springer.

US DoJ. 2020. *Six Russian GRU Officers Charged in Connection With Worldwide Deployment of Destructive Malware and Other Disruptive Actions in Cyberspace.* Press Release, October 19, 2020. Office of Public Affairs, U.S. Department of Justice. www.justice.gov/opa/pr/six-russian-gru-officers-charged-connection-worldwide-deployment-destructive-malware-and

Valeriano, B., and R.C. Maness. 2014. The Dynamics of Cyber Conflict Between Rival Antagonists, 2001–11. *Journal of Peace Research* 51 (3): 347–360.

van der Eijk, C. 2018. *The Essence of Politics*. Amsterdam University Press.

Van Puyvelde, D., and A. Brantly. 2019. Cybersecurity: Politics, Governance and Conflict in Cyberspace. *Polity*.

Vial, G. 2019. Understanding Digital Transformation: A Review and a Research Agenda. *The Journal of Strategic Information Systems* 28 (2): 118–144.

Virilio, P., and S. Lotringer. 1998. *Pure War*. Semiotext.

Von Solms, R., and J. Van Niekerk. 2013. From Information Security to Cyber-Security. *Computers & Security* 38 (October): 97–102.

Voo, Julia, Irfan Hemani, and Daniel Cassidy. 2022. *National Cyber Power Index 2022*. Belfer Center for Science and International Affairs, Harvard Kennedy School.

Waever, O. 2010. Towards a Political Sociology of Security Studies. *Security Dialogue* 41 (6): 649–658.

Walt, S.M. 1991. The Renaissance of Security Studies. *International Studies Quarterly* 35 (2): 211–239.

Warner, M. 2012. Cyber-security: A Pre-history. *Intelligence and National Security* 27 (5): 781–799.

Weiss, M., and V. Jankauskas. 2019. Securing Cyberspace: How States Design Governance Arrangements. *Governance* 32 (2): 259–275.

Whyte, C. 2018. Dissecting the Digital World: A Review of the Construction and Constitution of Cyber-Conflict Research. *International Studies Review* 20 (3): 520–532.

Whyte, C., and B. Mazanec. 2023. *Understanding Cyber-Warfare: Politics, Policy and Strategy*. 2nd ed. London: Routledge.

Wilson, J.Q. 1989. *Bureaucracy: What Government Agencies Do and Why They Do It*. Basic Books.

Winseck, D. 2019. Internet Infrastructure and the Persistent Myth of U.S. Hegemony. In *Information, Technology and Control in a Changing World*, ed. B. Haggart, K. Henne, and N. Tusikov, 93–120. Palgrave Macmillan.

Work, J.D. 2020. Evaluating Commercial Cyber-Intelligence Activity. *International Journal of Intelligence and Counter-Intelligence* 33 (2): 278–308.

Zetter, K. 2015. *Stuxnet and the Launch of the World's First Digital Weapon*. Crown.